Studies in Surface Science and Catalysis 12

METAL MICROSTRUCTURES IN ZEOLITES
Preparation — Properties — Applications

Studies in Surface Science and Catalysis

Studies in Surface Science and Catalysis 12

METAL MICROSTRUCTURES IN ZEOLITES

Preparation – Properties – Applications

Proceedings of a Workshop, Bremen, September 22–24, 1982

Editors

P.A. Jacobs
Centrum voor Oppervlaktescheikunde, Katholieke Universiteit Leuven, Leuven, Belgium

N.I. Jaeger
Forschungsgruppe Angewandte Katalyse, Universität Bremen, Bremen, F.R.G.

P. Jírů
Czechoslovak Academy of Sciences, J. Heyrovský Institute of Physical Chemistry and Electrochemistry, Prague, Czechoslovakia

G. Schulz-Ekloff
Forschungsgruppe Angewandte Katalyse, Universität Bremen, Bremen, F.R.G.

ELSEVIER SCIENTIFIC PUBLISHING COMPANY
Amsterdam – Oxford – New York   1982

ELSEVIER SCIENTIFIC PUBLISHING COMPANY
Molenwerf 1
P.O. Box 211, 1000 AE Amsterdam, The Netherlands

Distributors for the United States and Canada:

ELSEVIER SCIENCE PUBLISHING COMPANY INC.
52, Vanderbilt Avenue
New York, NY 10017

Library of Congress Cataloging in Publication Data
Main entry under title:

Metal microstructures in zeolites.

    (Studies in surface science and catalysis ; v. 12)
    Sponsored by Deutsche Forschungsgemeinschaft, and
others.
    Includes index.
    1. Zeolites--Congresses. 2. Microstructure--Congress-
es. I. Jacobs, Peter A. II. Deutsche Forschungsgemeinschaft (1951-   ) III. Series.
TP159.M6M47 1982   666'.86   82-13732
ISBN 0-444-42112-2 (U.S.)

ISBN 0-444-42112-2 (Vol. 12)
ISBN 0-444-41801-6 (Series)

Printed in The Netherlands

CONTENTS

VI

PREFACE

Well-defined microstructures in zeolites are becoming increasingly interesting as model systems for the study of structure-dependent processes on a molecular scale. An example for the clarification of a microstructure in a zeolite can be found in ultramarine. This zeolite has been used as a gem stone for more than five thousand years, ornamenting pharaohs and emperors, and has served as a paint and dye for centuries. Its preparation as paint was described by Marco Polo and its price was bewailed by Dürer. The presumable and fortuitous manufacture of lapis lazuli was reported by Goethe (Italienische Reise, Palermo, April 13, 1787), and its controlled preparation was achieved by Guimet on the basis of the analysis by Clément and Désormes. In the following decades it served as colour for the Meissener Porzellan. The Deutsches Reichspatent No. 1 (July 2, 1877) claims the preparation of ultramarine red. Only very recently could the structure of the sulphur-containing complexes within the channels of the ultramarine-type zeolite (which are responsible for the colour effect) be clarified by new chemical and physical methods, which points to the fact that we are only just beginning to understand model systems of this type. It can therefore be expected that new methods for the preparation and characterization of microstructures in zeolites will be helpful for the elucidation of physicochemical mechanisms, e.g. in catalysis or solar energy transfer, thus contributing to the development of economic processes in the conversion of materials and energy.

The Organising Committee are very grateful to the sponsors: Deutsche Forschungsgemeinschaft, Deutscher Akademischer Austauschdienst, Mobil Oil AG, and the University of Bremen, and appreciate their valuable contributions; we express our thanks to Mrs. B. Thomsen for her valuable editorial assistance; and we wish lively discussions for all participants.

THE CHEMISTRY OF RUTHENIUM IN ZEOLITES

J.H. Lunsford, Department of Chemistry,
Texas A&M University, College Station, Texas 77843 (U.S.A.)

ABSTRACT

Ruthenium may be exchanged into zeolites in the form of $[Ru^{III}(NH_3)_6]^{3+}$, $[Ru^{III}(NH_3)_5NO]^{3+}$, $[Ru^{II}(NH_3)_5N_2]^{2+}$, or $[Ru^{III}(H_2O)_5OH]^{2+}$ complexes. The hexa-ammine complex may be oxidized to the nitrosyl complex, which in turn may be reduced to the dinitrogen complex. In addition, $[Ru^{III}(NH_3)_6]^{3+}$ reacts with bipyridine to form $[Ru^{II}(bpy)_3]^{2+}$ within the cavities of a Y-type zeolite. The latter complex may be oxidized to Ru^{III} with Cl_2 and reduced back to Ru^{II} with H_2O. Several carbonyl complexes of ruthenium have been observed, some of which are important intermediates in the catalytic cycle for the water-gas shift reaction.

INTRODUCTION

The catalytic properties of ruthenium in zeolites [1-5] have prompted a number of studies on the chemistry of ruthenium complexes within the cavities of these crystalline aluminosilicates. The ammine and aquo complexes were viewed initially as a convenient means for achieving ruthenium ion exchange since "RuCl$_3$" is not suitable for this purpose; however, the reactions and properties of the complexes themselves are important in understanding the ultimate state of the ruthenium. Probably more is understood concerning the chemistry of ruthenium than any other metal ion in zeolites. This situation has arisen, in part, because of the various experimental probes which may be employed to study ruthenium.

The following discussion of ruthenium chemistry in zeolites is intended to be representative rather than comprehensive. Because of the problems associated with structural and stoichiometric determinations in zeolites, there is a certain amount of ambiguity in the identification of several complexes. Difficulties in the determining oxidation state, coordination number, etc. have sometimes given rise to disagreements in the literature. Other complexes exhibit spectra which are almost identical to those found in more conventional media, thus their identification is definitive. In some respects the former case is more desirable than the latter since often one would hope to synthesize in zeolites new complexes which have no analog in more conventional inorganic chemistry.

RUTHENIUM AMMINE, NITROSYL AND NITROGEN COMPLEXES

The most common starting material for the formation of Ru^0 particles is $[Ru^{III}(NH_3)_6]^{3+}$ which has been exchanged into a zeolite [1,2, 4-7]. The final state of the ruthenium depends very much on the atmosphere under which this complex is decomposed; that is, under H_2, O_2, a vacuum, or an inert atmosphere. Attempts have been made, therefore, to follow the intermediates in the decomposition process. First, it should be noted that in a Y-type zeolite the smallest metal particles may be obtained by heating a $[Ru^{III}(NH_3)_6]Y$ sample under vacuum followed by reduction in H_2 [1,7]. The fact that autoreduction by the ammonia ligand occurs was recognized by observing N_2 evolution which begins at about 300°C and by following the subsequent H_2 uptake during final reduction with hydrogen. Some disagreement exists, however, concerning the extent of this reduction. Elliott and Lunsford [1] report an autoreduction to an average oxidation state of Ru^{II}; whereas, Verdonck et al. [7] observed that up to 80% of the ruthenium was reduced to Ru^0. It is probably significant that less reduction occurred on samples which contained 0.5 and 2 wt% Ru while the 80% reduction occurred on a sample which contained 4 wt% Ru. Since Ru^0 is an effective catalyst for ammonia synthesis and decomposition, it is expected that formation of a few metal particles would greatly enhance the reduction process via the production of H_2.

The reduction of ruthenium during dehydration-deammination was followed by photoelectron spectroscopy as depicted in Fig. 1 [6]. The spectra are shown for samples which were degassed under vacuum at 25°C or progressively heated to higher temperatures in flowing He. Evacuation at 25°C or heating in He up to 115°C gave poorly resolved ruthenium lines, although in the latter case a shoulder having a binding energy of ca. 283.0 eV was observed. At 300°C there was a shift in binding energy to 281.7 eV, and raising the temperature to 400°C caused a further shift to 281.4 eV. The binding energy shifted to 281.0 eV upon reduction in H_2. Although the interpretation of these results in complicated by the fact that the metal particle size has an effect upon the binding energy, the most straightforward interpretation is that the shift to lower binding energies upon treatment in H_2 is due to a change in oxidation state of the ruthenium.

Even at 25°C $[Ru^{III}(NH_3)_6]^{3+}$ undergoes a reaction both in air and under vacuum as indicated by a change in color of the zeolite from nearly white to pink and then to a deep wine-red. This color was first attributed to $[Ru^{III}(NH_3)_5OH]^{2+}$ [8], but Madhusudhen et al. [9] have correctly assigned it to ruthenium-red $(NH_3)_5Ru^{III}-O-Ru^{IV}(NH_3)_4-O-Ru^{III}(NH_3)_5$. Although the diffuse reflectance spectrum is dominated by this complex (λ_{max} = 780,550,390,258 nm) its concentration in the zeolite may be rather low because of a large extinc-

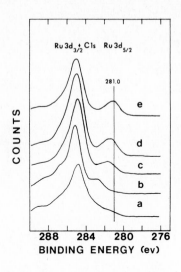

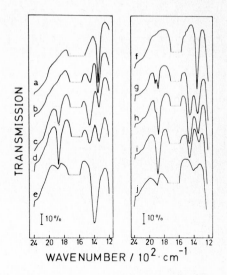

Fig. 1. XPS spectra of Ru3d$_{5/2}$ and Ru3d$_{3/2}$ + C1s levels of 2% RuNaY (a) after evacuation overnight at 25°C, (b) after degassing in flowing He in increments of 100°C/hr to 115°C, (c) to 300°C, (d) to 400°C, (e) after treatment as in (d) followed by reduction in flowing H$_2$ for 16-18 hr at 400°C [6].

Fig. 2. Infrared spectra of selected RuX and RuY zeolites: (a)[Ru^{III}(NH$_3$)$_6$]X starting zeolite; (b-e) heated in flowing O$_2$ to 70, 110, 177 and 257°C; (f) [Ru^{III}(NH$_3$)$_6$]Y starting zeolite; (g-j) heated in flowing O$_2$ to 70, 110, 177 and 257°C. The band at 1650cm^{-1} is not shown for clarity [11].

tion coefficient. The ruthenium-red complex is essentially destroyed by heating the sample under vacuum or O$_2$ at 125°C [10,11].

Upon heating the [Ru^{III}(NH$_3$)$_6$]-zeolite (X or Y) to 70°C in O$_2$ an infrared band was observed at 1925 cm^{-1} which has been attributed to [Ru^{III}(NH$_3$)$_5$NO]$^{3+}$ or perhaps [Ru(NH$_3$)$_4$(NO)(OH$_2$)]$^{3+}$. Under flowing O$_2$ at 180°C a strong infrared band was observed at 1870-1875 cm^{-1}, and it is also evident from Fig. 2 that a band at 1335 cm^{-1} due to coordinate NH$_3$ remained. The data suggest that [Ru(O$_{zeol}$)$_3$(NH$_3$)NO] was formed [11], which has been more generally described as a [Ru(NO)(NH$_3$)$_{1,2}$(O$_1$,OH,H$_2$O)$_{4,3}$] [10].

Infrared spectra show that evacuation up to 300°C is required to eliminate all of the coordinated NH$_3$ and NO [10]. When the sample was heated in O$_2$ at 200°C, the [Ru(O$_{zeol}$)$_3$(NH$_3$)NO] complex was destroyed and RuO$_2$ was formed on the external surface of the zeolite, as indicated by the large increase in the XPS band of ruthenium [6].

Nitrosyl complexes may be introduced directly into zeolites by ion exchange of [Ru^{III}(NH$_3$)$_5$NO]Y [8]. The spectrum of this complex shown in Fig. 3a exhibits bands at 1927 cm^{-1} (νNO), 1640 cm^{-1} (δNH$_3$ + residual H$_2$O) and 1365 cm^{-1} (NH$_3$, sym). This is a much more thermally stable complex than [Ru^{III}(NH$_3$)$_6$]$^{3+}$

in either X or Y zeolites. When NO was adsorbed on freshly prepared [Ru^{III}(NH$_3$)$_6$]Y zeolite a strong infrared band at 1918 cm^{-1} was observed and attributed to the [Ru^{III}(NH$_3$)$_5$NO]$^{3+}$ complex [10].

The [Ru^{III}(NH$_3$)$_5$NO]$^{3+}$ complex in zeolite Y undergoes several interesting reactions to form dinitrogen complexes of ruthenium. The reaction of [Ru^{III}(NH$_3$)$_6$]$^{3+}$ with NO, described above, resulted in a band at 2120cm^{-1} which is characteristic of [Ru(NH$_3$)$_5$N$_2$]$^{2+}$. Verdonck et al. [10] proposed the following mechanism for this reaction:

$$[Ru^{III}(NH_3)_6]^{3+} + NO \rightarrow [Ru(NH_3)_5(NH_2NOH)]^{3+} \tag{1}$$

$$[Ru(NH_3)_5(NH_2NOH)]^{3+} + NH_3 \rightarrow [Ru(NH_3)_5N_2]^{2+} + NH_4^+ + H_2O \tag{2}$$

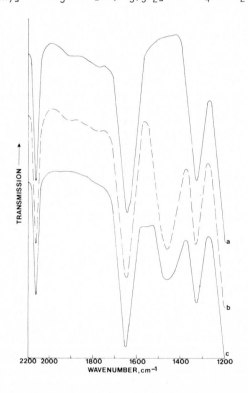

Fig. 3. Infrared spectra of [Ru^{III}(NH$_3$)$_5$NO] zeolite (a) after degassing over-night at 25°C, (b) after the addition of N$_2$H$_4$ (7.5 Torr), and (c) degassing for 45 min [8].

In addition, the exposure of a [Ru^{III}(NH$_3$)$_5$NO]Y zeolite to vapor phase hydra-zine gave rise in a very strong infrared absorption due to dinitrogen at 2132 cm^{-1}, as depicted in Fig. 3b and c [8]. Before evacuation bands were observed at 1885 cm^{-1}, and at 1460 cm^{-1}. The former is attributed to

$[Ru(NH_3)_5(NO)(NHNH_2)]^{2+}$ and the latter to NH_4^+, obtained from the decomposition of hydrazine. After degassing the sample only the spectrum of the $[Ru(NH_3)_5N_2]^{2+}$ complex remained. In the absence of moisture and oxidizing agents this complex was stable for several months. Direct exchange of $[Ru^{II}(NH_3)_5N_2]^{2+}$ into the zeolite was possible, and the dinitrogen band at 2132 cm^{-1} was observed [8]. In the presence of the zeolitic water the complex was very unstable and was rapidly transformed into ruthenium-red. Laing et al. [8] carried out experiments to determine whether the $[Ru^{II}(NH_3)_5N_2]^{2+}$ complex may activate N_2 for the synthesis of ammonia. Upon addition of N_2 and H_2 to the zeolite NH_3 was formed; however, isotopic labelling experiments showed that the ammonia was derived from residual N_2H_2 rather than from molecular nitrogen.

RUTHENIUM BIPYRIDINE COMPLEXES

Tris(2,2'-bipyridine)ruthenium(II) has been used in several schemes for the photodecomposition of water [12,13]. In an effort to explore the influence of zeolites on the photochemical and photophysical properties of transition metal complexes $[Ru^{II}(bpy)_3]^{2+}$ was synthesized within a Y-type zeolite. This synthesis involved the reaction of $[Ru^{II}(NH_3)_6]Y$ zeolite with solid bipyridine (mixed in a 4:1 or 3:1 bpy:Ru mole ratio) at 200°C [14,15]. At a loading of 9.1 cations per unit cell approximately 75% of the ruthenium reacted to form complexes. Photoelectron spectroscopy was used to confirm that the complexes were formed within the zeolite cavities rather than on the external surface. The diameter of the complex is 10.8 Å. The $[Ru^{II}(bpy)_3]^{2+}$ complex in zeolite Y is characterized by a maximum in the diffuse reflectance spectra at 460 nm with a shoulder at 430 nm (Fig. 4.) Two bands in the 425-454 nm region have been reported by several investigators for the $[Ru^{II}(bpy)_3]^{2+}$ complex in different media [16,17]. The complex in the orange-colored zeolite exhibits a luminescence spectrum at 590 nm which is in good agreement with λ_{max} in aqueous media [18].

One of the most dramatic effects in the zeolite was the strong inverse relationship between emission intensity and concentration, which is characteristic of self-quenching or concentration quenching. This phenomenon is the result of resonance transfer of electronic excitation energy from an initially absorbing molecule to another identical molecule until the quanta reaches a quenching site [19]. In the zeolite a quenching site may be a $[Ru^{II}(bpy)_3]^{2+}$ complex adjacent to a transition metal impurity. Concentration quenching is particularly apparent in the zeolites because of the high complex concentration; e.g. concentrations of 10^{-5} M are typical in aqueous solution studies whereas for a $[Ru^{II}(bpy)_3]Y$ (6.9) sample the effective concentration is 0.77 M. For the more dilute zeolite samples a luminescence lifetime of 579 ns was

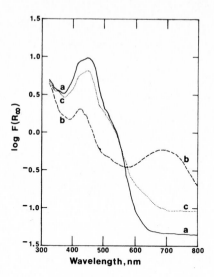

Fig. 4. Visible-region diffuse-reflectance spectra: (a) outgassed [Ru^{II}(bpy)$_3$]Y; (b) outgassed [Ru^{III}(bpy)$_3$]Y; (c) sample b, after exposure to water [15].

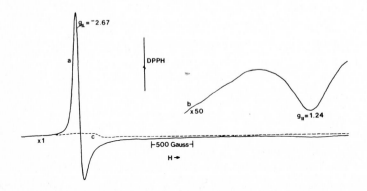

Fig. 5. Electron paramagnetic resonance spectra taken at -196°C: (a) [Ru^{III}(bpy)$_3$]Y; (b) sample a, at 50 X greater receiver gain; (c) sample a, following exposure to water and outgassing [15].

comparable to that reported for dilute solutions of [Ru^{II}(bpy)$_3$]$^{2+}$ in oxygen-free H$_2$O [17,20].

The effect of water adsorbed in the zeolite was ultimately to quench emission, although in a dilute zeolite sample, having 2.8 complexes per unit cell, quantities of water up to 80% of saturation actually enhanced the emission [14]. The origin of this enhancement is uncertain; however, it may be the result of increasing the distance between a ruthenium complex and a quenching ion through hydration of one or both ions. Since the transfer probability is proportional to the inverse sixth power of the distance between complex and quencher, increasing this distance by a few percent would cause a significant increase in emission. Luminescence from $[Ru^{II}(bpy)_3]^{2+}$ was also quenched by by the presence of molecular oxygen. Such quenching, which is well documented in solution studies [21], leads to the formation of singlet oxygen and the superoxide ion.

A half-reaction for the photodissociation of water involves the reduction of H_2O, or perhaps more correctly OH^-, by the $[Ru^{III}(bpy)_3]^{3+}$ complex. In order to test this half-reaction in a zeolite Ru(II) in the bipyridine complex was oxidized to Ru(III) with Cl_2 [15]. Evidence for the oxidation of 90% of the Ru^{II} was obtained from the diffuse reflectance spectra of Fig.4. Here the absorption maxima at 685 and 420 nm (curve b) are in good agreement with literature values of 680 and 418 nm. Moreover, $[Ru^{III}(bpy)_3]^{3+}$ exhibits an epr spectrum (Fig. 5) which is characterized by $g_\perp = -2.67$ and $g_{||} = 1.24$. (There is some ambiguity concerning the negative sign of $g_\perp$, but this choice was made for purpose of comparison with the existing literature.) Addition of water to the $[Ru^{III}(bpy)_3]$ zeolite indeed affected the reduction of the Ru(III), as idicated by curve c of Fig. 4, but no oxygen was detected in the gas phase. Carbon dioxide in an amount equivalent to about 6% of the total ruthenium was observed and labelling experiments using $H_2^{18}O$ confirmed that the oxygen in the CO_2 was derived from the H_2O. The immobility of the Ru(III) complex in the zeolite, as well as an unfavorable local chemical environment, probably inhibits the secondary electron-transfer process necessary for O_2 formation.

RUTHENIUM AQUO AND HYDROXY COMPLEXES

For various purposes it would be desirable to have ruthenium in a single oxidation state, coordinated only to the oxide ions of the zeolite. With many of the first row transition metal ions this objective may be easily achieved since the aquo complexes are obtained simply by dissolving in water a salt of the desired cation. The hexaaquo complex of Ru(III), by contrast, is relatively difficult to synthesize, and it exists in this form only at low pH values with weakly coordinating anions.

Coughlan et al. [22] previously reported the preparation of $[Ru^{III}(H_2O)_6]^{3+}$ and its exchange into several zeolites, but the mode of preparation of the complex and the diffuse reflectance spectra of the zeolites make it questionable as to whether this proposed complex was actually introduced into the zeolite. Most significant is the fact that the exchange was carried out at a pH of 6.5 where $[Ru^{III}(H_2O)_6]^{3+}$ is unstable and polymerized hydroxy species are known to occur [23].

In our laboratory $[Ru^{III}(H_2O)_6]^{3+}$ was prepared in 2 M HBF_4 using the method of Kallen and Early [24]. The presence of fluoride ions proved to be deleterious to the zeolite crystallinity; therefore, a method suggested by Buckley and Mercer [25] was used to substitute toluene sulfonate ions for tetrafluoroborate ions via an anion exchange resin. Optical spectra of the solution revealed bands at λ_{max} = 290 nm, characteristic of $[Ru^{III}(H_2O)_5(OH)]^{2+}$, and at λ_{max} = 392 nm, characteristic of $[Ru^{III}(H_2O)_6]^{3+}$ [23,26]. As the pH was increased by adding NaOH, the band at 390 nm decreased and the band at 295 nm increased. At pH values greater than 2 it was observed that λ_{max} shifted to longer wavelengths, until at pH of 4.3 λ_{max} = 325, which may be indicative of a $[Ru^{III}(H_2O)_4(OH)_2]^+$ complex.

Ion exchange of the ruthenium complexes was carried out at pH = 3.8 [26]. X-ray diffraction spectra of the air-dried sample confirmed that the crystallinity of the zeolite was maintained, provided toluene sulfonate was the anion. The epr spectrum of the zeolite after degassing at 25°C for 16 hr is shown in Fig. 6. The spectrum is characterized by $g_\perp$ = 2.426 and $g_{\parallel}$ = 1.714

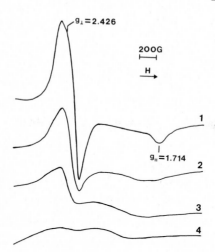

Fig. 6 Epr spectra of $[Ru^{III}(H_2O)_5OH]Y$ zeolite (1) after degassing 16 hr at 25°C; (2) after degassing 1 hr at 100°C; (3) after degassing 2 hr at 200°C; (4) after degassing 3 hr at 300°C [26].

which are similar to the values of $g_\perp$ = 2.373 and $g_{\parallel}$ = 1.722 obtained for $[Ru^{III}(H_2O)_5OH]^{2+}$ in an aqueous solution [26]. Since under the pH conditions of the exchange reaction $[Ru^{III}(H_2O)_5OH]^{2+}$ was the major component of the solution, it is expected that this monomeric species was present within the zeolite. It is interesting to note from curve 2 of Fig. 6 that more than 50%

of these hydroxy complexes remained even after most of the water had been removed from the zeolite by heating the sample under vacuum at 100°C. Presumably the ruthenium remains in the III oxidation state upon complete decomposition of the complex, and the decrease in epr intensity is due to a change in spin state, rather than to a reduction in oxidation state. In contrast to bipyridine ligands it is expected that oxide ions of the zeolite would stabilize Ru(III) in the presence of water.

RUTHENIUM CARBONYL AND OXYGEN COMPLEXES

Ruthenium in zeolites forms several carbonyl complexes; however, the oxidation state of the metal in these complexes is not well defined. The nature of the complexes is particularly important because of the activity of ruthenium in methanation, Fischer-Tropsch and water-gas shift reactions [1-5]. Verdonck et al. [3] have shown that a RuX zeolite is the most active known catalyst for the water-gas shift reaction at 90°C, and the exceptionally high activity of this catalyst is maintained up to ca. 290°C, at which temperature ruthenium is reduced to the metallic state. An active RuY catalyst exhibited intense infrared bands at 1960, 2020 and 2085 cm^{-1}. The bands at 2020 and 2085 cm^{-1} were attributed to a dicarbonyl complex of Ru(I) and the band at 1960 to a monocarbonyl complex of Ru(II):

$$Ru(I) \begin{matrix} \diagup CO \\ | \quad \diagdown CO \\ NH_3 \end{matrix} \qquad and \qquad Ru(II) \begin{matrix} \diagup NH_3 \\ \diagdown CO \end{matrix}$$

Both species are present in an active water gas shift catalyst.

Goodwin and Naccache [27] have reported that a rich infrared spectrum was observed when CO was adsorbed on a well reduced RuY catalyst. The spectra shown in Fig. 7 have been interpreted in terms of surface complexes: species A having CO vibrational bands at 2156 and 2100 cm^{-1} with perhaps a shoulder at 2086 cm^{-1}; species B having a band at 2075 cm^{-1}; and species C having its main band at 2045 cm^{-1} and possibly possessing bands at 2134 and 1934 cm^{-1}. Several possible assignments have been given for species A; however, the most reasonable is an unreduced form of Ru(II) having multiply adsorbed CO. Species B is interpreted as a metal-cluster ruthenium carbonyl in which the metal is in a zero valence state, and species C is believed to be a metal particle with adsorbed CO. In comparing these results with those of Verdonck et al. [3] it is surprising that the ionic forms of ruthenium in structures 1 and 2 would have stretching frequencies at such low wavenumbers compared to those of zero valence ruthenium.

Paramagnetic carbonyl complexes of ruthenium have been observed by epr

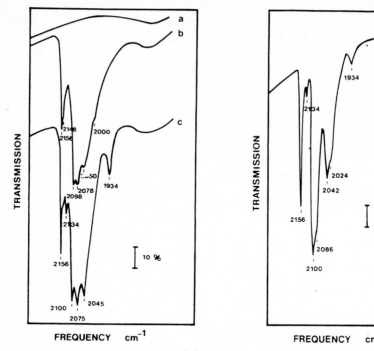

Fig. 7. Infrared spectra of RuY (a) following decomposition of the catalyst in flowing H_2 at 404°C and desorption at 404°C; (b) after adsorption of CO from 155 Torr CO at 25°C for 17 hr; (c) after adsorption of CO from 190 Torr CO at 150°C for 2 hr; (d) after exposure to 149 Torr O_2 at 25°C for 141 hr [27].

spectrscopy following decomposition of a $[Ru^{III}(NH_3)_6]Y$ zeolite at 300°C and exposure to CO [28]. Upon exposure to excess CO (100 Torr) at 25°C the epr spectrum of Fig. 8, curve a, was observed. The species giving rise to the spectrum is believed to be a Ru(III) polycarbonyl complex. Upon evacuation ofgas-phase CO another spectrum with $g_\perp = 2.060$ and $g_\parallel = 1.9871$ was observed as shown in curve c. This spectrum was attributed to a Ru(III) monocarbonyl complex. Spin concentration calculations indicate that the observable paramagnetic Ru-CO species accounted for only 1-2% of the total ruthenium present in the zeolite. Hydrogen reduction at 300°C for 19 hr resulted in complete loss of observable paramagnetic ruthenium complexes.

Adsorption of O_2 onto a sample which contained the monocarbonyl complex resulted in an epr spectrum characteristic of a superoxide ion. The carbonyl spectrum of Fig. 8 c reversibly disappeared and a new spectrum having $g_1 = 2.001$, $g_2 = 2.006$ and $g_3 = 2.083$ was observed. Oxygen-17 hyperfine splitting indicates that the two oxygens of the superoxide ion were nonequivalent and that the electron was almost completely localized in the π-antibonding orbitals on the oxygen. An equilibrium reaction of the type

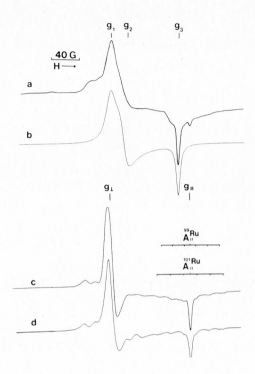

Fig. 8. Epr spectra of ^{12}CO adsorbed on dehydrated RuY zeolite: (a) 100 Torr of CO added at 25°C; (b) computer simulation of (a); (c) 100 Torr of CO at 25°C followed by evacuation at 25°C to $<10^{-3}$ Torr; (d) computer simulation of (c) [28].

was proposed in which the metal was effectively oxidized to Ru^{IV}.

With respect to the effect of oxygen on ruthenium-exchanged zeolites it is important to note that RuO$_2$ is readily formed at 25°C from small metal parti-cles or from ionic forms of ruthenium at temperatures in excess of 300°C. Ruthenium dioxide is highly mobile and at elevated temperatures will rapidly migrate to the external surface of the zeolite where crystals on the order of 500Å are formed [7]. Rapid exposure of a sample containing small Ru^0 particles to air at 25°C will result in migration of some RuO_2 to the external surface as the oxidation reaction causes local heating within the zeolite [6]. An undesirable consequence of this phenomenon is that one cannot obtain an oxidized form of ruthenium <u>in a zeolite</u> simply by heating the material in O_2 to elevated temperatures.

CONCLUSIONS

Clearly ruthenium in zeolites has a rich and varied chemistry which in many respects parallels its behavior in homogeneous media. The complexes that have been studied confirm that ruthenium forms strong nitrosyl complexes, volatile oxides, and weak aquo complexes. The photophysical properties of $Ru(bpy)_3$ complexes in zeolites are similar to those found in aqueous solution, but the photochemical properties differ primarily because of restricted motion and the absence of certain free ions (e.g. OH^-). A variety of carbonyl complexes may be formed with ruthenium in several valence states, and these complexes may play a role in the interesting catalytic chemistry of carbon monoxide.

REFERENCES

1 D.J. Elliott and J.H. Lunsford, J. Catal.,57 (1979) 11.
2 B.L. Gustafson and J.H. Lunsford, J. Catal., in press.
3 J.J. Verdonck, P.A. Jacobs and J.B. Uytterhoeven, J.C.S. Chem. Commun. 1974) 181; J.J. Verdonck, R.A. Schoonheydt and P.A. Jocobs, Proc. 7th Int. Congr. Catalysis, Tokyo, June 30-July 4, 1980, Kodansha Ltd., Tokyo, 1981, pp. 911-924.
4 H.H. Nijs and P.A. Jacobs, J. Catal., 65 (1980) 328.
5 P.A. Jacobs, in B. Imelik (Ed.), Catalysis by Zeolites, Elsevier, Amsterdam, 1980, pp. 293-308.
6 L.A. Pedersen and J.H. Lunsford, J. Catal., 61 (1980) 39.
7 J.J. Verdonck, P.A. Jacobs, M. Genet and G. Poncelet, J.C.S. Faraday I, 76 (1980) 403.
8 K.R. Laing, R.L. Leubner and J.H. Lunsford, Inorg. Chem., 14 (1975) 1400.
9 C.P. Madhusudhan, M.D. Patil and M.L. Good, Inorg. Chem., 18 (1979) 2384.
10 J.J. Verdonck, R.A. Schoonheydt and P.A. Jacobs, J. Phys. Chem., 85 (1981) 2393.
11 J.R. Pearce, B.L. Gustafson and J.H. Lunsford, Inorg. Chem., 20 (1981) 2957.
12 J. Kiwi, E. Borgarello, E. Pelizzetti, M. Visca and M. Grätzel, Angew. Chem. Int. Ed. Engl., 19 (1980) 647.
13 J.-M. Lehn, J.-P. Sauvage, R. Ziessel, Nouv. J. Chim., 4 (1980) 355.
14 W. DeWilde, G. Peeters and J.H. Lunsford, J. Phys. Chem., 84 (1980) 2306.
15 W.H. Quayle and J.H. Lunsford, Inorg. Chem., 21 (1982) 97.
16 F.E. Lytle and D.M. Hercules, J. Am. Chem. Soc., 91 (1969) 253.
17 J. Van Houten and P.J. Watts, J. Am. Chem. Soc., 98 (1976) 4853.
18 F.E. Lytle and D.M. Hercules, J. Am. Chem. Soc., 88 (1966) 4745.
19 D.L. Dexter and J.H. Schulman, J. Chem. Phys., 22 (1954) 1063.

20 C.-T. Lin, W. Böttcher, M. Chou, C. Creutz and N. Sutin, J. Am. Chem., 98 (1976) 6536.
21 V.S. Srinivasan, D. Podolski, N.J. Westrick and D.C. Neckers, J. Am. Chem. Soc. 100 (1978) 6513.
22 B.S.Coughlan and W.A. McCann, Chem. Ind., 19 (1976) 527.
23 Z. Harzion and G. Navon, Inorg. Chem., 19 (1976) 2236.
24 T.W.Kallen and J.E. Earley, Inorg. Chem., 10 (1971) 1149.
25 E.E. Mercer and R.R. Buckley, Inorg. Chem., 4 (1965) 1692.
26 B.-Z. Wan and J.H. Lunsford, Inorg. Chim. Acta, in press.
27 J.G. Goodwin and C. Naccache, J. Catal., 64 (1980) 482.
28 B.L. Gustafson, M.-J. Lin and J.H. Lunsford, J. Phys. Chem., 84 (1980) 3211.

EPR STUDIES OF Co(II)-BIS(DIMETHYLGLYOXIMATO)-COMPLEXES AND THEIR OXYGEN ADDUCTS IN A ZEOLITE X MATRIX

C.J. Winscom*, W. Lubitz**
* Institut für Molekülphysik, ** Institut für Organische Chemie,
Freie Universität Berlin, D-1000 Berlin 33
H. Diegruber and R. Möseler[†]
Forschungsgruppe Angewandte Katalyse, Fachbereich 3 - Chemie,
Universität Bremen, D-2800 Bremen 33
[†] Present address: Grace GmbH, D-6520 Worms

ABSTRACT

Bis(dimethylglyoximato)-Co(II) complexes have been synthesized in a cobalt-exchanged NaX zeolite. The electronic structures of a sequence of samples derived from this complex by infusion of water and oxygen were studied by EPR spectroscopy. The g- and ^{59}Co hyperfine tensors of the individual species were determined and compared with model systems. The zeolite occluded species could be assigned and a plausible reaction scheme proposed.

INTRODUCTION

Bis(dimethylglyoximato)-(dmgH)$_2$-complexes are known to catalyse a variety of organic reactions. Species with coordinated molecular hydrogen, oxygen, and even olefins have been postulated as reactive intermediates. Co(dmgH)$_2$ complexes have been of particular interest in view of the understanding of bonding and activation of oxygen in biological and catalytic systems (ref. 1). The synthesis of such complexes within the supercages of a zeolite offers several advantages over "normal solutions": i) the complexes are immobilised, ii) dimerisation processes of the complexes are retarded and iii) the zeolite might stabilise the complexes by acting as a ligand, thereby altering the electronic structure and its catalytic activity. Previous studies of Co(II) complexes with nitrogen chelating ligands have been reported by other authors (refs. 2, 3) using EPR, IR and optical reflectance spectroscopy. In this work we have performed a simple hydration and oxygenation sequence with a view to assigning all the species observed by EPR techniques, and thereby constructing a plausible reaction scheme and a theoretical rationalisation.

EXPERIMENTAL

X-band EPR measurements were performed on a Bruker ER 200 D spectrometer. A 100 kHz field modulation was used; the modulation depth was typically 5 G_{pp}. All spectra were recorded at 120 K.

Co(II)X zeolites with different Co contents were prepared from highly-purified self-synthesised NaX faujasite (Si : Al = 1.23) by conventional ion-exchange. These were designated $Co_{2.2}NaX$, $Co_{7.3}NaX$ and $Co_{15.1}NaX$; the respective subscripts indicate the number of Co ions per unit cell. Prior to ion-exchange, the NaX was heated under shallow-bed conditions (air flow, 2 K/min to 673 K, 673 K for 10 h) to remove organic contaminants. After cooling to room temperature, the NaX was rehydrated by placing it over water in a closed vessel.

The Co-exchanged faujasites were dehydrated by heating (purified Ar flow, 1-2 K/min to 673 K, 673 K for 5 h). Co_nNaX and dried $dmgH_2$ were sealed in Ar under reduced pressure (10^{-3} Pa) in the individual compartments of a two chamber reaction vessel. The $dmgH_2$ was then allowed to infuse into the Co_nNaX (373 K, 40 h), with frequent shaking of the zeolite bed to ensure homogeneity (sample A). Part of this sample was hydrated (10^3 Pa water vapour, 5 h). Excess water was then removed under reduced pressure (1 Pa, 2 h) (sample B). The hydrated sample was then exposed to oxygen (10^5 Pa, 293 K, 5 h) (sample C).

RESULTS

I. Model complexes

The primary aim of this work has been the characterisation of the different species present in zeolite samples A, B and C. We have found it useful to refer to two model systems, namely Co(II) $(dmgH)_2$ diluted in Ni(II) $(dmgH)_2$ as a co-crystallised powder, and Co(II) $(dmgH)_2$ in a frozen water solution. The first system has an axial geometry which precludes effective axial coordination effects. We take this as a model for an almost pure four-coordinate planar complex. In the second system, the solvent provides a large excess of a weakly coordinating donor ligand. In frozen solution we take this as a model for six-fold coordination, in which the two axial positions are each occupied by a water molecule (cf. ref. 4). The EPR spectra of both systems are typical of randomly-

oriented doublet (S = 1/2) ground state species and are characterised by the principal values of the g- and ^{59}Co hyperfine (hf)
tensors, which have been determined by fitting computer simulated
spectra to those observed. The values obtained are presented in
Table 1, and are consistent with a half-filled orbital having
primarily d_{z^2} character (z = out-of-plane axis).

II. Zeolite samples

Co_nNaX, prior to infusion of $dmgH_2$, exhibits a broad EPR
spectrum in the range g ≈ 6 to g ≈ 2, consistent with Co^{2+} in a
near tetrahedral crystal field of four nearest neighbour oxygen
atoms (ref. 5). After infusion of $dmgH_2$ into Co_nNaX (sample A) an
additional EPR signal, representing ca. 20 % of the total Co^{2+}
present, is observed in the g ≈ 2 region. This region is shown
in Fig. 1, together with a fitted simulation; the g- and ^{59}Co hf
principal values are listed in Table 1 as species 1.

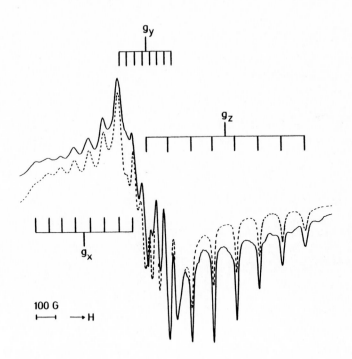

Fig. 1.   The EPR spectrum of sample A (prepared from $Co_{7.3}$NaX)
resulting mainly from species 1 (—— Experimental, --- Simulated)

Subsequent infusion of water to form sample B results in the
spectrum shown in Fig. 2. It is several times more intense than
species 1 in sample A; the Co^{2+} in the tetrahedral zeolite environment almost disappears. In addition to species 1, the spectrum
in Fig. 2 contains two further species (2 and 3). These are identified with the help of results from sample C (species 2) and
samples, described elsewhere (ref. 8), in which a greater proport-
ion of infused H_2O is present (species 3).

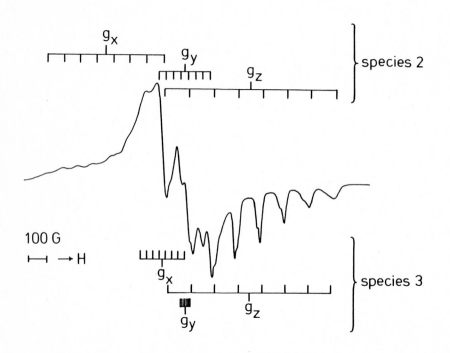

Fig. 2.   The EPR spectrum of sample B (derived from $Co_{7.3}NaX$). The
principal features for species 2 and 3 are as indicated; additional features in the spectrum arise from species 1.

After exposure of sample B to oxygen to form sample C, two
easily distinguished species, shown in Fig. 3, are observed. The
first may be characterised and shown to be one of the components
(species 2) in B. The second (species 4) is much narrower, having
almost isotropic g- and ^{59}Co hf-tensors. The principal hf values

are markedly reduced in magnitude compared with species 1, 2 and
3. The parameters for species 4 are typical of super-oxo cobalt
complexes (ref. 6). The spin Hamiltonian parameters for species 2,
3 and 4 are also listed in Table 1.

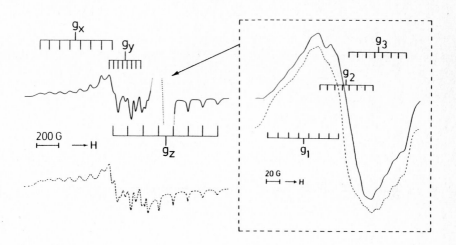

Fig. 3. The EPR spectrum of sample C (derived from $Co_{7.3}NaX$).
The broader spectrum arises from species 2 (—— Experimental,
--- Simulated). The narrower superposed spectrum, detailed in the
insert, is that of species 4 (—— Experimental, --- Simulated).

DISCUSSION

   The similarity of species 2 with the model six-coordinate complex and also species 3 with the model four-coordinate complex is
readily seen from Table 1. Species 1 has parameters intermediate
between the four- and six-coordinate models. With the help of an
analysis of spin Hamiltonian parameters (ref. 7) and simple
assumptions about the crystal field (ref. 8), the parameters for
a five-coordinate (axial H_2O) complex may be estimated by inter-
polation. Species 1 is then seen to have parameters lying broadly
between the four- and five-coordinate cases. Potential fifth
ligands are oxygen at the wall of the zeolite cage or water possibly
formed in the reaction. The nature of species 1 leads us to believe
that the relevant Co to oxygen distance is larger than that ex-
perienced in the free water complex (six-coordinate model species).
We have found similarities in the behaviour of the $Co(dmgH)_2$-

TABLE 1

The experimentally determined principal values of the g- and
[59]Co hf tensors for the various model compounds and zeolite species.
g-value errors are ±0.001 and for the hf values typically ±3 MHz.
With the exception of species 4, all species are consistent with a
$(d_{z^2})^1$ ground state in which all principal hf values are taken to
have a positive sign.

	Co(dmgH)$_2$ in Ni(dmgH)$_2$	Co(dmgH)$_2$ in H$_2$O	Species 1 Samples A,B	Species 2 Samples B,C	Species 3 Sample B	Species 4 Sample C
g_x	2.595	2.310	2.543	2.605	2.312	(g_1)2.052
g_y	2.245	2.194	2.262	2.196	2.190	(g_2)2.015
g_z	1.995	2.009	2.003	1.980	1.980	(g_3)1.986
A_x(MHz)	370	70	252	336	126	(A_1) 43
A_y (")	185	20	114	132	27	(A_2) 33
A_z (")	378	328	315	348	338	(A_3) 36

species in the zeolite and that of the known normal solution
chemistry of this complex (ref. 9). One exception to the solution
chemistry behaviour is that no noticeable depletion of the paramagnetic Co species occurs as a result of dimer formation, e.g. the
diamagnetic μ-oxo complex (ref. 6), since the dimensions of the
zeolite α-cages hinder the presence of more than one Co-complex.
Samples derived from Co$_{15.1}$NaX, which, after hydration would yield
approximately two Co^{2+} complexes per α-cage, were studied to check
this possibility.

The sequence A → B → C can be rationalised as follows: infusion
of dmgH$_2$ allows complexation of mainly those Co^{2+} ions already
present in the α-cages, resulting in species 1. Subsequent infusion
of water aids the release of Co^{2+} from "hidden" sites, allowing
ultimate access to the uncomplexed dmgH$_2$. In this sample we believe
that species 2,3 and probably 1 exist in equilibrium with each
other. We anticipate that an excess of water in sample B would
produce almost exclusively a six-coordinate complex (species 3).
Finally, a super-oxo-complex is formed from the water carrying
species. Increasing water coordination destabilises the half-
filled d$_{z^2}$ orbital relative to d$_{xz}$, d$_{yz}$. At some stage its energy
will exceed that of an unoccupied π* molecular orbital of O$_2$, and

the formation of the $Co(III)O_2^-$ adduct is energetically favoured.

OUTLOOK

We are currently trying to clarify open questions concerning the participation of species 1, 2 and 3 in the expected equilibria by altering the preparative conditions. The short-comings of X-band EPR (anisotropic linebroadening in randomly oriented species) have prompted us to examine alternative magnetic resonance approaches. Of particular interest in this respect are: i) W-band (90 GHz) EPR, ii) zero-field EPR and iii) conventional X-band ENDOR spectroscopy.

ACKNOWLEDGEMENTS

We thank Dr. A. Schweiger (ETH, Zürich), and Dr. M. Plato and M. Baumgarten (both F.U. Berlin) for providing spectra of model compounds and computational assistance. We are grateful for many useful discussions with Prof. N. Jaeger, Dr. P. Plath and Prof. G. Schulz-Ekloff (Universität Bremen). Financial support by the DFG (SFB 161) is gratefully acknowledged.

REFERENCES

1 G.N. Schrauzer and L.P. Lee, J.Am.Chem.Soc. 92 (1970) 1551.
2 R.F. Howe and J.H. Lunsford, J.Am.Chem.Soc. 97 (1975) 5156, J.Phys.Chem. 79 (1975) 1836.
3 R.A. Schoonheydt and J. Pelgrims, J.C.S. Dalton (1981) 914.
4 E.K. Ivanova, I.N. Marov, A.T. Panfilov, O.M. Petrukhin, and V.V. Zhukov, Zhur.neorg.Khim. 18 (1973) 1298.
5 I.D. Mikheikin, G.M. Zhidomirov, and V.B. Kazanskii, Russ. Chem.Rev. 41 (1972) 468.
6 R.D. Jones, D.A. Summerville, and F. Basolo, Chem.Rev. 79 (1979) 139.
7 B.R. McGarvey, Can.J.Chem. 53 (1975) 2498.
8 W. Lubitz, C.J. Winscom, R. Möseler, and H. Diegruber, to be published.
9 G.N. Schrauzer and R.J. Windgassen, Chem.Ber. 99 (1966) 602.

IN-SITU UV/VIS-MICROSCOPESPECTROPHOTOMETRY OF REACTIONS OF Co^{2+}-COMPLEXES WITHIN SINGLE CRYSTALS OF FAUJASITE-TYPE ZEOLITES

H. DIEGRUBER and P.J. PLATH

Forschungsgruppe Angewandte Katalyse, Universität Bremen, Bibliothekstraße, D-2800 Bremen 33 (Germany)

ABSTRACT

The dehydration of CoNaX zeolite single crystals has been observed "in-situ" by UV/VIS microcomputer controlled microscopespectrophotometry (MMSP) depending upon the temperature in successive limited steps from room temperature up to $350^{\circ}C$.

In addition the MMSP-method has been used to observe the in-situ formation of Co^{2+}-complexes with dimethylglyoxime ($dmgH_2$) within the zeolite cavities, the $dmgH_2$ entering from the gas phase at a given temperature.

The structural changes the Co^{2+}-complexes in the zeolite undergo during the dehydration and adsorption of dimethylglyoxime are discussed and the validity of the method is demonstrated.

INTRODUCTION

UV/VIS-spectra of Co^{2+} exchanged zeolites are well-known in the literature (ref. 1-6). Scientists are interested in the electronic structure of the complexes formed by Co^{2+} ions within the zeolite framework in order to get some information concerning their role in catalytic processes. The investigations are usually carried out by reflection spectroscopy using well-prepared zeolite powders. Spectra of hydrated and dehydrated zeolites as well as spectra of highly complex compounds such as cobalt-ethylenediamine within the zeolite cavities have been reported (ref. 7).

These UV/VIS-spectra may give a good overview regarding the state of the sample under a given set of conditions, there is however little information with respect to the real situations during dehydration processes or the formation of the complexes. We therefore intended to investigate the processes by in-situ-UV/VIS measurement at temperatures at which they take place. It seems

that the method of UV/VIS-transmission microscope-spectrophotome-
try on single crystals is just the right method for solving our
problem (ref. 8,9).

METHODS

 CoNaX zeolites with different cobalt content were prepared by
conventional ion exchange using NaX single crystals synthesized
following the method described by Charnell (ref. 10).

 These are designated $Co_{7.2}NaX$, $Co_{13.6}NaX$ and $Co_{17.3}NaX$ where
the subscripts indicate the number of cobalt ions per unit cell.

 The transmission spectra of single crystal zeolites were recorded with a microscope-spectrophotometer UVMP (Leitz) in the ranges
of 230-410 nm and 370-800 nm with different lamps using a procedure described previously in more detail (ref. 8). Each spectral region has been scanned by a stepping motor driven by a microcompute
in 1000 steps with 1 or 10 fold accumulation of the spectra, respectively. For the dehydration of CoNaX zeolite single crystals
the temperature has been varied linearly ($5^oC/min$) in successive
limited steps of about 25^oC from room temperature up to 350^oC.
During this time a constant flow of argon (99,997 %, 120 ml/min)
dried on molecular sieves, passed the zeolite crystal situated in
a special cell in the optical path of the microscope. The transmission spectra were recorded 3 to 5 minutes after the single crystal has reached the chosen temperature. Each measurement took a
time of about 6 min. Thereafter the temperature was raised again.

 Prior to the adsorption the $dmgH_2$ was dried in a vacuum for several hours within the cell. Then the $Co_{7.2}NaX$ crystals dehydrated
externally were placed near the $dmgH_2$ crystals. To follow the formation of Co^{2+}dimethylglyoximato-complexes the cell was evacuated
again to about 1.3 Pa and the crystals located within the range of
the optical path were heated rapidly (10 or $20^oC/min$) to a chosen
temperature (max. 100^oC) and kept there for a given time. The process of diffusion of $dmgH_2$ into the zeolite crystal is a relatively fast process compared with the time needed for recording the
spectra. In order to follow the formation of Co^{2+}-complexes with
$dmgH_2$ within the zeolite cavities our method had to be changed:

 For the measurement the selected zeolite crystals were cooled
suddenly to room temperature. An additional diffusion of $dmgH_2$ into the zeolite is neglectable during the whole time of the spectro-

scopic measurement of about 10 minutes. Such a long time is needed since a tenfold accumulation of the spectra was required. After each spectroscopic measurement the temperature was raised again.

RESULTS

As a starting point for further investigations the spectra of hydrated CoNaX crystals with different cobalt content per unit cell ($Co_{7.2}NaX$, $Co_{13.6}NaX$ and $Co_{17.3}NaX$) were recorded at room temperature (see Fig. 1). For the lowest degree of exchange the typical

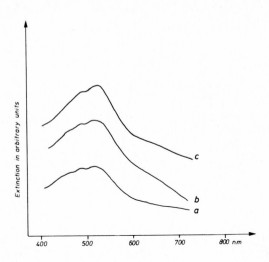

Fig. 1. Absorption spectra of hydrated cobalt exchanged faujasite single crystals. (a) $Co_{7.2}NaX$; (b) $Co_{13.6}NaX$; (c) $Co_{17,3}NaX$

spectra of a rhombic distorted octahedral Co^{2+}-complex with three absorption maxima at 471, 491 and 521 nm are obtained due to the Jahn-Teller effect. For higher degrees of ion exchange only two bands of this triple remain at 490 and 520 nm, and a shoulder at about 690 nm arises at the same time indicating the transition to an tetragonal distorted octahedral ligand field.

It is obvious that Co^{2+} ions in zeolite X prefer arrangements of low symmetry even in the hydrated framework. This behavior of Co^{2+} ions is similar to that of the Ni^{2+} ions in comparable situations (ref. 9).

The dehydration process was studied using crystals of $Co_{7.2}NaX$ (see Fig. 2). During the temperature increase the spectrum changes dramatically. Even for low temperatures we get the typical spectra

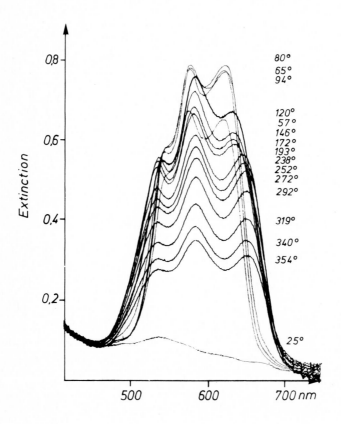

Fig. 2. Dehydration of a $Co_{7.2}NaX$ zeolite single crystal (after one hour in an argon flow).

of tetrahedral Co^{2+}-complexes marked by the band quadruple between 450 and 700 nm due to $^4F(\Gamma_2) \rightarrow {}^4P(\Gamma_4)$ transition, which splits up due to (L,S) coupling effects (ref. 11). In a pure cubic field at most 4 of the 12 possible independent states of ^4P should be observable. Spectra like this can be found for crystals of $(n\text{-}Bu_4N)_2$ $(CoCl_4)$ for example (ref. 12). But in general one has to assume a trigonal distortion to complexes with symmetry C_{3v} by solvation. As a result only three bands can be observed like we did at higher temperatures.

Fig. 3 and 4 depict that several periods have to be distinguished during dehydration: first the period of the formation of a

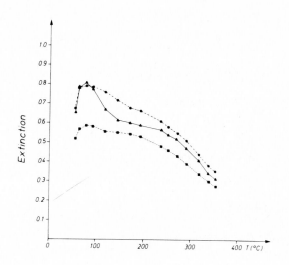

Fig. 3. The extinction of main transition bands of a $Co_7.2NaX$ single crystal during dehydration.

tetrahedral complex. This symmetry is characterized by four bands at 486, 548, 577 and 622 nm. Continuing the dehydration process by increasing the temperature we observe the formation of a complex with symmetry C_{3v} instead of Td. The symmetries may be assigned to the formation of a free moving tetrahedral complex within the cavities in the first period and an immobilized trigonal C_{3v}-complex during further dehydration.

In Fig. 4 the shift of band maxima can be followed. This can be observed for all degrees of exchange in tendency.

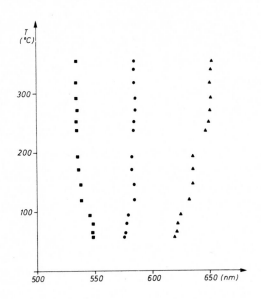

Fig. 4. Dependence of the position of main transmission bands on dehydration of a $Co_{7.2}NaX$ single crystal.

For a detailed discussion of this process further results concerning the temperature dependence of the extinction coefficients as well as an analysis of zeolites with different contents of Co^{2+} ions are needed. A more detailed discussion will be presented elsewhere (ref. 13).

Here we like to present some results of our investigations of the adsorption process of $dmgH_2$ on $Co_{7.2}NaX$ instead.

Fig. 5 shows the spectra of the starting-point and the final state of the adsorption of $dmgH_2$ on a totally dehydrated $Co_{7.2}NaX$ zeolite crystal. The spectrum of the final state is characterized by a broad and strong band in the UV-region, which extends into the visible range and exhibits the threefold band splitting typi-

cal for distorted tetrahedrally coordinated complexes. To prove
the presumption that there are hidden bands among the unstructured
flank of the UV-band, the adsorption process was followed by in-
situ measurements (see Fig. 6).

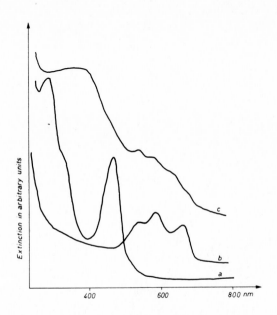

Fig. 5. Complete spectra of (a) $Co(dmgH)_2 \cdot 2H_2O$ (solvent: water);
(b) $Co_{7.2}NaX$ dehydrated (Ar, 400°C, 5h), (c) $Co_{7.2}NaX$ dehydrated
(Ar, 400°C, 5h) and adsorption of $dmgH_2$ at 100°C.

Keeping the system at 100°C for successive short time respectively, we can follow the growth of a double band at 423/439 nm
eventually becoming a shoulder at the visible flank of the UV-band.
A "splitting" of the red band of the C_{3v} band tripel is connected
with this process.

If we assume the formation of the $(Co(dmgH)_2(O_{zeolite})_2$-complex
within the zeolite we have to base our arguments on a D_{4h}-symmetry,
which is a subgroup of O_h. For this reason we should observe the
reverse process of the previous one: the transition of $C_{3v}(T_d)$-
symmetric complexes to complexes with $D_{4h}(O_h)$ symmetry. As a re-

sult the spectral structure should be the same in principle as in the case of the hydrated form. Fig. 5 shows the spectra of the aqueous solution of $Co(dmgH)_2 \cdot 2H_2O$ for comparison. The formation of the "blue" double-band can now be identified with the original $\Gamma_4(^4F) \rightarrow \Gamma_4(^4P)$ transition in the octahedral field, or the $\Gamma_2(^4F) \rightarrow \Gamma_5(^4P)$ and $\Gamma_2(^4P)$ transition in cases of D_{4h}-symmetry respectively.

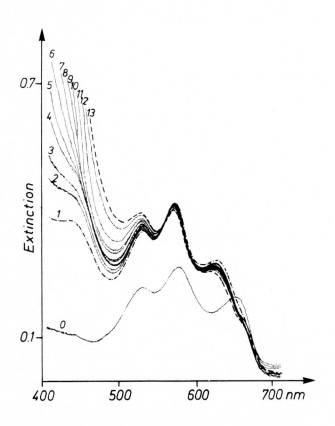

Fig. 6. Adsorption of $dmgH_2$ by $Co_{7.2}NaX$ dehydrated ($400^{\circ}C$, Ar 5h), (0 before adsorption, 1-13 adsorption at $100^{\circ}C$).

There is another fact, we have to take care of: the increase of intensity of the tetrahedral band system as well as the splitting of the red band. Fig. 7 shows a detailed picture of the very begin of the adsorption process giving us the possibility to discuss the single adsorption steps more precisely.

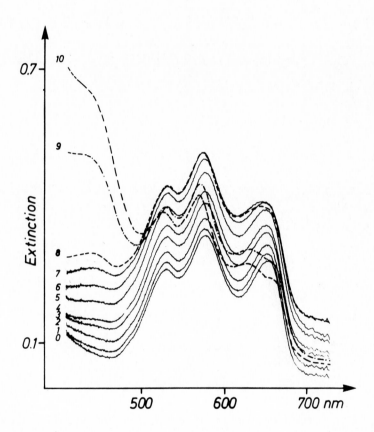

Fig. 7. First adsorption steps of dmgH$_2$ by Co$_{7,2}$NaX dehydrated (380°C, Ar 5h), (0 before adsorption; 1-6, 7-8, 9-10 adsorption at 55°C, 65°C, 100°C respectively).

If the covalency of the bonding in a tetrahedral complex is increased an increase of the extinction should be observed as a consequence (ref. 14). This argument in mind we remember the structure of the dmgH$_2$. For this molecule there will be a possibility to form compounds analogous to metal-nitrosyl complexes. Such ligands are much stronger than (OH)$^-$ or H$_2$O ligands. Comparing our results with the statements by Lunsford et al. (ref. 15) for tetrahedral cobalt-nitrosyl complexes in zeolites, we find nearly the same spectral data (see Fig. 7). Lunsford observed the shift of the red band from 15270 cm^{-1} (654,9 nm) to 15630 cm^{-1} (639,8 nm) during the formation of nitrosyl complexes. This leads to the conclusion that the first step of the adsorption of dmgH$_2$ is connected with the formation of a C$_{3v}$-distorted tetrahedral nitrosyl-like Co^{2+}-complex. In a second reaction step the formation of the classical D$_{4h}$-distorted octahedral complex can be observed. In our spectra the original Co^{2+}-complex can still be observed following the reaction with dmgH$_2$ since there remains a band at 655 nm, which did not vanish. Only about 25% of this complex has reacted during several hours at 100°C.

SUMMARY

Following the dehydration process of Co$_{7.2}$NaX zeolites and following the adsorption of dmgH$_2$ by UV/VIS spectroscopic in-situ measurements on single crystals we obtained detailed informations about the sequence of symmetries of the ligand partition.

To discuss the location of these complexes with respect to the zeolite framework would be highly speculative, even though it would be very tempting to do so. It is our opinion that we do not gain any significant information by trying to express our results in terms of migration paths and sites within the zeolite.

REFERENCES

1   K. Klier, Adv. Chem. Ser. 101 (1971), 480
2   B. Wichterlova, P. Jiru and A. Curinova, Z. Phys. Chem. (Frankfurt a.M.) 88 (1974), 80
3   M.A. Heilbron and J.C. Vickermann, J. Catal. 33 (1974), 434
4   H. Hoser, S. Kryzanowski and F. Trifiro, J. Chem. Soc., Faraday Trans. 1,71 (1975), 665
5   P.J. Hutta and J.H. Lunsford, J.Chem. Phys. 66,10 (1977),4716
6   H. Praliand and G. Coudurier, J. Chem. Soc., Faraday Trans. 1,75 (1979), 2601
7   R.A. Schoonheydt and J. Pelgrims, J.C.S. Dalton 4 (1981), 914
8   E.C. Hass, P.J. Plath and H. Weigel, Leitz Mitt. Wiss. u. Techn. 7 (1980), 220
9   E.C. Hass and P.J. Plath, J. Mol. Catal.14 (1982), 35
10 J.F. Charnell, J. Cryst. Growth 8 (1971), 291
11 C.J. Ballhausen and C.K. Jørgensen, Acta Chem. Scand. 9 (1955) 397
12 F.A. Cotton, D.M.L. Goodgame and M. Goodgame, J. Am. Chem. Soc. 83 (1961), 4690
13 H. Diegruber, K. Koblitz and P.J. Plath, in preparation
14 F.A.Cotton and R.H.Soderberg,J.Am.Chem.Soc. 89 (1962), 872
15 J.H. Lunsford, P.J. Hutta, M.J. Lin and K.A. Windhorst, Inorg. Chem. 17 (1978), 606

MOSSBAUER SPECTROSCOPIC STUDIES OF FERROUS ION EXCHANGE IN ZEOLITE A

L.V.C. REES

Physical Chemistry Laboratories, Imperial College of Science and Technology, London SW7 2AY.

ABSTRACT

Mossbauer spectroscopy has been used to determine the optimum conditions required to produce the maximum degree of ferrous exchange in the sodium form of zeolite A with a minimum in the degree of hydrolysis. On dehydration two doublets are observed in the spectra. The low quadrupole split doublet is associated with ferrous ions located in site I, trigonally coordinated to three nearest-neighbour framework oxygens. The outer quadrupole split doublet could be produced by ferrous ions sited in site I coordinated to three framework oxygens and a near-neighbour framework hydroxyl group. However, by comparing the spectra obtained from dehydrated ferrous zeolite A containing sodium, calcium and ammonium co-ions it has been concluded that the outer doublet is associated with ferrous ions located in collapsed framework material.

INTRODUCTION

Mossbauer spectroscopy of Fe(57) zeolites has given information on the location, chemical state and effect of sorbate molecule interaction on the iron cations present in the zeolites. The exchange of iron cations into the less acid stable zeolites is not easily accomplished and often leads to spurious components in the spectra due to hydroxide species precipitated on the external surfaces of the zeolite or to some degree of lattice collapse. In this paper a more detailed study of the factors which affect the exchange of ferrous ions into zeolite A will be described along with a brief survey of the properties of these ferrous ions.

EXPERIMENTAL

The zeolite was shaken with 15% sodium chloride solution for 24 h at room temperature to ensure that no other cation or entrained salt was present. The zeolite was then washed with distilled water at pH 7 until no trace of chloride could be found in the washings. This starting material and the ferrous exchanged samples were chemically analysed using the method of Begel'fer et al [1]. Water in the zeolite was determined by thermal gravimetry and silicon by difference. The unit cell formula of the starting material was found to be $Na_{12}Al_{12}Si_{12}O_{48} \cdot 26 \cdot 9H_2O$.

The crystallinity of all the zeolite samples used in these studies was determined from oxygen sorption capacities at liquid nitrogen temperatures. Under an equilibrium pressure of 14.49 k Pa the oxygen uptake of the starting material was 192 cm^3 at s.t.p. g^{-1} (of dehydrated zeolite).

The ferrous exchanged zeolite samples were transferred as a slurry in distilled water into a Pyrex glass Mossbauer sample cell under oxygen free, nitrogen gas in the usual way [2]. The glass cell had two very thin Pyrex windows separated by a 1 mm gap. The sample filled this gap after the excess water was removed at room temperature under vacuum. It is essential to keep the dry zeolite sample free from contact with air or oxygen as some immediate oxidation to the ferric form occurs.

The Mossbauer spectrometer has been fully described elsewhere [2].

RESULTS

Experiments were first carried out to determine the optimum conditions to produce the maximum degree of ferrous exchange in zeolite A with a minimum in the degree of hydrolysis. The results of these experiments are summarized in Tables 1 and 2.

TABLE 1

Ferrous exchange in zeolite A at pH 5.

Sample	Initial pH	Wt. of Fe $SO_4.7H_2O$ (g)	Exchange time (h)	Unit cell formula of product
1	5.0	0.5	0.5	$Na_{4.88} H_{5.58} Fe_{0.77}$ [A]
2	5.0	2.0	0.5	$Na_{4.15} H_{5.75} Fe_{1.05}$ [A]
3	5.0	2.0	1.0	$Na_{3.10} H_{5.56} Fe_{1.67}$ [A]
4	5.0	3.0	1.0	$Na_{1.76} H_{7.16} Fe_{1.54}$ [A]
5	5.0	4.0	1.5	$Na_{1.65} H_{7.67} Fe_{1.34}$ [A]

TABLE 2
Ferrous exchange in zeolite A : effect of initial pH

Sample	Molality	Acid cm^3	Initial pH	Unit cell formula of product
1	0.125	10.0	4.3	$Na_{1.12} H_{9.80} Fe_{0.54}$ [A]
2	0.125	7.5	4.7	$Na_{2.39} H_{7.45} Fe_{1.08}$ [A]
3	0.125	5.0	5.0	$Na_{3.07} H_{5.81} Fe_{1.56}$ [A]
4	0.050	9.0	6.0	$Na_{4.47} H_{4.99} Fe_{1.27}$ [A]
5	0.0125	12.0	6.5	$Na_{3.61} H_{2.72} Fe_{2.83}$ [A]
6	0.0125	5.0	8.0	$Na_{3.67} H_{1.59} Fe_{3.37}$ [A]

[A] represents $Al_{12}Si_{12}O_{48} \cdot xH_2O$

In these experiments the initial pH was adjusted to the required value by
adding 0.125M H_2SO_4 in the case of samples in Table 1, or the quoted
molality in the case of samples in Table 2, to a slurry of 0.5 g of Na-A in
45 cm^3 of distilled water over a period of 20 min. The required weight of
solid ferrous sulphate hydrate plus 1 mg of ascorbic acid was then added over
a period of 5 min. Exchange was then allowed to proceed for the required
time. Agitation was maintained by bubbling oxygen free, nitrogen gas through
the reaction mixture. In the experiments summarized in Table 2, 2g of ferrous
salt and one hour exchange time was used in all cases. All experiments were
carried out in a glove box under an oxygen free, nitrogen atmosphere.

Table 1 samples 2 and 3 shows that increasing the exchange time from
half an hour to one hour increases the ferrous content significantly but not
the degree of hydrolysis. Increased ferrous sulphate concentration and time
of exchange in samples 4 and 5 actually decreased the ferrous content of the
zeolite and increased the degree of hydrolysis. In Table 2 it is shown that
an increase in the initial pH decreases the degree of hydrolysis and increases
the ferrous content of the zeolite. Figure 1 shows that a decrease in the
H^+ content is correlated with an increase in the Fe^{2+} content of the zeolite.

These experiments suggest that the conditions used to obtain sample 6
Table 2 would seem to give the optimum exchange assuming that no precipitation
of hydrous iron oxide has occurred on the external surfaces of the zeolite
crystals. Monitoring of the pH of the slurry during the exchange indicated
that such precipitation was unlikely. The pH was found to decrease continuously

on adding the ferrous sulphate and ascorbic acid e.g. in experiments 4, 5 and 6 in Table 2 the pH dropped to 5.0, 5.2 and 5.6 respectively.

The room temperature Mossbauer spectra of samples 3, 5 and 6, Table 2 are shown in Figure 2. All three spectra have similar broadened doublets with isomer shifts (IS) of 1.20 - 1.22 mm s^{-1} (with respect to natural iron), quadrupole splittings (QS) of 2.06 - 2.12 mm s^{-1} and linewidths, Γ, of 0.33 - 0.36 mm s^{-1}. Because of the high recoil free fraction, f, associated with these hydrated zeolite samples it is concluded that the spectra in Figure 2 are produced, therefore, by Fe^{2+} ions located in site 1, six-fold coordinated to three framework oxygens of the 6-ring and three water molecules in the α cage. There is no evidence of any other type of Fe^{2+} species being present which confirms experiments on the thermal stability of zeolite A containing appreciable quantities of H^{+} ions that significant partial lattice collapse only occurs on dehydration of these acid forms at elevated temperatures (> 100°C).

When the same samples were now dehydrated at 100°C under vacuum the spectra, taken at room temperature, shown in Figure 3 were obtained. Each spectrum now exhibits two doublets arising from two different types of coordination around the Fe^{2+} ions. The parameters of the outer doublet (IS = 1.13 - 1.07 mm s^{-1}, QS = 2.30 - 2.12 mm s^{-1}) show a small decrease in IS and a small increase in QS compared with the parameters found above for the completely hydrated samples. These results are consistent with Fe^{2+} ions located in site I but now co-ordinated to three (or possibly fewer) water molecules in a more distorted geometry than exists in the fully hydrated zeolites. However, these parameters could also be produced by Fe^{2+} ions located in collapsed framework "sites". The low QS doublet is produced by Fe^{2+} ions located in site I trigonally co-ordinated to three framework oxygens. Even after mild 100°C dehydration some 70 - 80% of the Fe^{2+} ions have no water ligands.

When these samples were now dehydrated at 360°C under vacuum for 24 h the spectra given in Figure 4 show that the area under the inner doublet increased at the expense of the area under the outer doublet in the case of samples 5 and 6 but the reverse occurred for sample 3. On continuing the dehydration at 360°C for a further 48 h, in an attempt to remove the last traces of water in the channels, the area under the inner doublets increased at the expense of the area under the corresponding outer doublets in all three samples. The parameters of the outer doublet (IS = 1.09 - 0.98 mm s^{-1}, QS = 2.42 mm s^{-1}) show a further small decrease in the IS and small increase in the QS on heating to this higher temperature.

When sample 3 was dehydrated at $360^{\circ}C$ under vacuum for total times of 120 and 168 h little change in the parameters were found although a very slight increase in the area of outer doublet at the expense of the inner doublet was observed.

In all of these samples the parameters of the inner doublet stayed sensibly constant (IS = 0.83 - 0.84 mm s^{-1}, QS = 0.46 - 0.48 mm s^{-1}) indicating that any small decrease in the degree of dehydration or small amount of lattice collapse had no affect on those trigonally coordinated Fe^{2+} ions sited in site I.

After dehydration at $360^{\circ}C$ for 24 h or longer all three samples gave spectra in which the percentage area of the outer doublet showed some correlation with the degree of hydrolysis given in Table 2. With sample 6 the correlation was exact. With sample 5 the correlation was excellent after the 24 h dehydration but after 72 h dehydration the percentage area of the outer doublet was much smaller than the degree of hydrolysis. With sample 3 the outer doublet percentage area decreased from 36 to 29% on increasing the outgassing time from 24 to 72 h but the area increased again to 34% on continued heating for 168 h. The percentage area, however, was always much less than the degree of hydrolysis of 48%.

The outer doublet of these samples would seem to be associated with Fe^{2+} ions contained in collapsed framework material. However, the outer doublet in samples 5 and 6 may be produced by Fe^{2+} associated with framework OH groups produced in the hydrolysis reaction. Dehydroxylation of these OH grups is not likely to have taken place at $360^{\circ}C$ even under vacuum. The outer doublet would not seem to be associated with precipitated hydrous iron oxide on the surface of the zeolite crystals as it is unlikely that any correlation between the degree of cation deficiency and the amount of oxide precipitated would be found.

An attempt was made to prove that the outer doublet of sample 6 was associated with OH groups on the framework. This sample was heated to $500^{\circ}C$, the maximum temperature the Mossbauer cells could sustain under vacuum. The percentage area of the outer doublet remained constant at 14-15% suggesting that OH groups were not responsible for the outer doublet. However, dehydroxylation often requires temperatures in excess of $550^{\circ}C$ and this conclusion needs further verification.

Calcium ions are known to increase the thermal stability of zeolite A. Ferrous ions were introduced into a zeolite A, which had some of the sodium ions replaced with calcium ions, under conditions which were exactly analogous to those used in the preparation of sample 6, Table 2. The resulting zeolite had a unit cell composition of $Fe_{1.61} Ca_{2.65} Na_{2.30} H_{1.18}[Al_{12} Si_{12} O_{48}] \cdot x H_2O$.

Figure 5 shows the spectra obtained from the fully hydrated Fe, Ca-A zeolite recorded at temperatures from ambient to 100K. All of these spectra can be fitted as two pairs of nested doublets. At room temperature the IS and QS values of the outer doublet are 1.24 and 2.36 mm s^{-1} and at 100K 1.37 and 3.08 mm s^{-1} respectively while the corresponding values for the inner doublet are 1.24 and 1.75 and 1.37 and 2.14 mm s^{-1} respectively. The ferrous ions responsible for the outer peaks are six-fold coordinated to three water molecules and three framework oxygen atoms in site I. The inner-peak ferrous ions are also located in site I and six-fold coordinated but one or more of the water ligands are replaced by hydroxyl groups. This asymetry in the ligands surrounding the ferrous ions produces the lower QS value. The larger electric field gradient (efg) resulting from the replacement of the univalent sodium ions by divalent calcium ions seems to encourage either the production of hydroxyl groups and/or to slow the proton jump process. In the case of ferrous exchanged hydrated Na-A the appearance of two nested doublets only occurred when the sample was cooled to temperatures much lower than ambient [3].

The spectra of the Fe, Ca-A zeolite were taken at different stages of dehydration (see Figure 6). Complete dehydration was accomplished on out-gassing the zeolite at 360°C for 24 h. The presence of calcium ions necessitated the use of higher temperatures to accomplish complete dehydration compared with those required for the Fe, Na-A zeolites. The IS of 0.82 mm s^{-1} and QS of 0.49 mm s^{-1} at room temperature agree well with the parameters obtained for this ferrous species in the dehydrated Fe, Na-A samples described previously. The replacement of sodium ions by calcium ions does not seem to have any influence on the parameters of the trigonally coordinated ferrous ions. In the calcium exchanged sample there is no evidence of the outer broad doublet found with the sodium exchanged samples. The unit cell formula of the Fe,Ca-A indicated the presence of $\sim$ 10% H^{+}. If the outer doublet found in the Fe,Na-A zeolites was due to the interaction of OH groups with the ferrous ions then one would expect to observe a similar outer doublet in the spectrum of dehydrated Fe, Ca-A. The lack of such a doublet would seem to confirm that this doublet arises from ferrous ions sited in collapsed framework material and provides further evidence of the increased thermal stability of calcium exchanged forms.

In a final set of experiments the spectra of Fe, NH$_4$-A zeolite were obtained after outgassing at increasing temperatures. Ammonium exchanged zeolite A is very unstable at temperatures where deammoniation and dehydroxylation occurs. The area under the outer doublet was now a large fraction of the total spectral area and a reasonable correlation between the degree of lattice collapse and

the area of this doublet was found. The parameters found for this doublet
did not change significantly on increasing the outgassing temperature through
the range where dehydroxylation was taking place. These results confirm,
once again, that the outer doublet is associated with ferrous ions sited in
collapsed lattice framework rather than some interaction with OH groups in
framework locations very close to the ferrous ions sited in site I.

REFERENCES
1 K.I. Begel'fer, P.A. Sazonova and K.M. Funtikova, Steklo i Keram,
   4 (1962) 30.
2 F.R. Fitch, Ph.D. Thesis, University of London, 1979.
3 B.L. Dickson and L.V.C. Rees, J. Chem. Soc. Faraday Trans 1,
   70 (1974) 2038.

APPENDIX: FIGURES

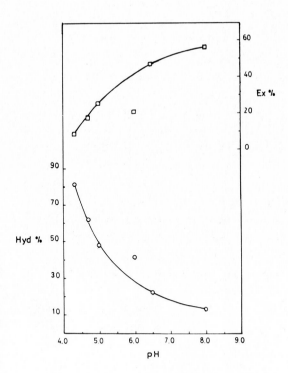

Fig. 1. Correlation between degree of hydrolysis and ferrous exchange.

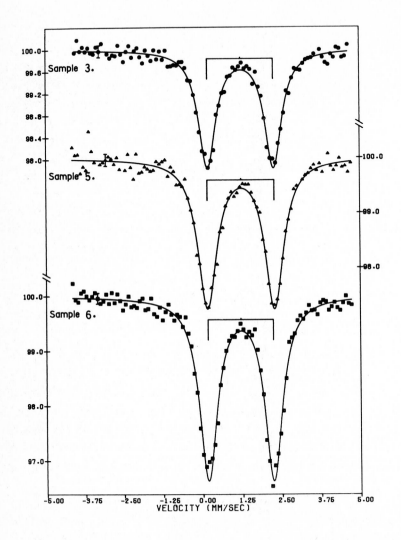

Fig. 2. Room temperature Mossbauer spectra of completely hydrated Fe, Na-A zeolites.

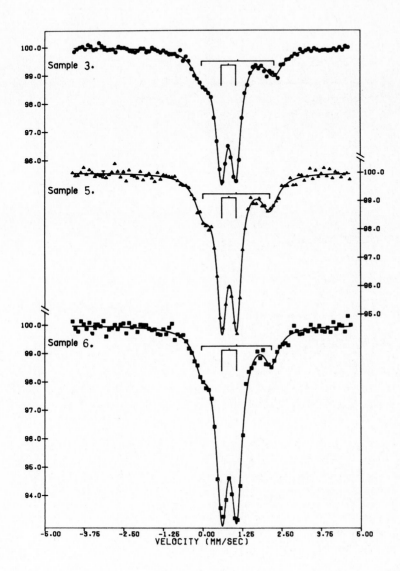

Fig. 3. Room temperature Mossbauer spectra of Fe, Na-A zeolites dehydrated at 100°C.

42

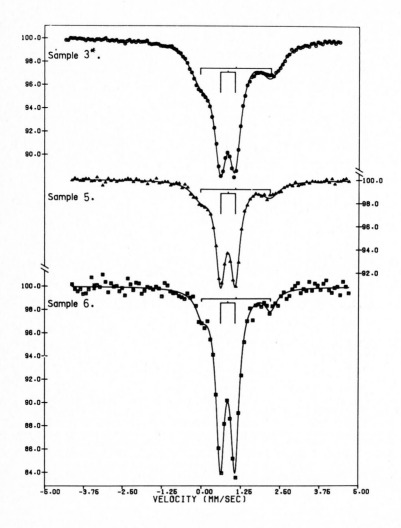

Fig. 4. Room temperature Mossbauer spectra of Fe, Na-A zeolites dehydrated at 360°C for 24 hours.

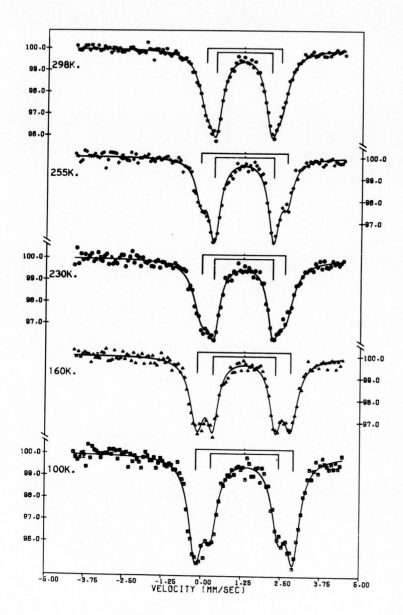

Fig. 5. Mossbauer spectra of completely hydrated Fe, Ca-A zeolite recorded at various temperatures.

44

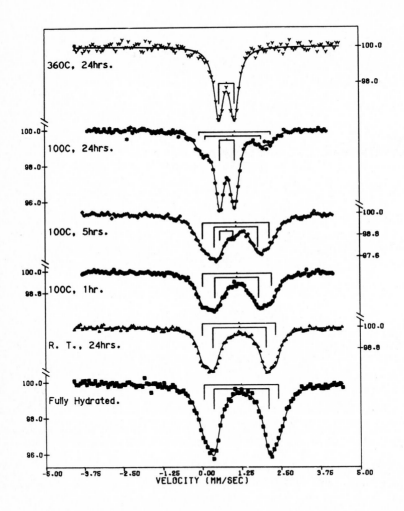

Fig. 6. Room temperature Mossbauer spectra of Fe, Ca-A zeolite following various outgassing treatments.

METAL INCLUSION COMPLEXES OF ZEOLITE A

N. PETRANOVIĆ[1] and R. DIMITRIJEVIĆ[2]
[1]Institute of Physical Chemistry, Faculty of Science, Beograd
(Yugoslavia)
[2]Faculty of Geology and Mining, Beograd (Yugoslavia)

ABSTRACT

By the process of denitration in vacuum at $520^{\circ}C$, a nitrate inclusion complex of zeolite $Ag_{12}A \cdot 9.5AgNO_3 \cdot 5.9H_2O$ is transformed into a metal inclusion complex. Analysis of X-ray diffraction powder data of the metal inclusion complex show that the lattice is cubic F-centered. The unit cell $a_0 = 2.4626(2)$ nm and space group, Fm3c, have been determined by single crystal X-ray diffraction techniques.

INTRODUCTION

By using melt as a reaction medium it is possible to replace water molecules in channels and cages of zeolite by molecules of salt. In that way inclusion complexes of zeolite are formed.

The nitrate inclusion complexes that are resulting from the contact of zeolite 4A and nitrate melts ($LiNO_3$ and $AgNO_3$), are: $Li_{12}A \cdot 9.8LiNO_3 \cdot 9.3H_2O$ and $Ag_{12}A \cdot 9.5AgNO_3 \cdot 5.9H_2O$ (A is the aluminosilicate part of the framework). These are complexes of cubic structure with the lattice constant $a_0 = 1.2075(1)$ nm and $a_0 = 1.2340(1)$ nm, respectively (ref. 1). In this way, the concentration of metal ions in cages is considerably increased, and the structure analysis gives the positions of counter ions and included cations. Besides the position 8(g) that the monovalent ions occupy in a unit cell of 4A zeolite, there are some new cation positions: 12j and 6f in alpha cage and 6e in beta cage in space group Pm3m. Most of the cations in alpha cage are situated towards the cage walls forming a "monolayer" of metal ions. These ions are in coordination, on one hand, with nitrate groups situated towards the centre of alpha cage, and, on the other hand, with the oxygens of the framework.

The highest density of the cations is placed in alpha and beta cages at a D4R ring, which is particularly evident in the case of lithium inclusion complex, and is manifested by TO_4 and D4R bands splitting in the IR spectrum of zeolite inclusion complex.

The aim of this work was the transformation of the metal nitrate inclusion complex into the metal inclusion complex, using the process of denitration at the temperature of decomposition of the included nitrates.

EXPERIMENTAL

The preparation of nitrate inclusion complexes is described elsewhere (ref. 1,2).

A Du Pont 1090 Thermal Analyzer was employed with High Temperature Cell 1200°C and TGA 951 Analyzer. The heating rate was 20°/min.

IR spectra were recorded at room temperature in the range 2000 - 250 cm^{-1} on a Perkin-Elmer grating spectrophotometer, type 457, using the KBr technique.

X-ray Diffractograms were obtained with a PHILIPS, PW-1051, powder diffractometer using a Ni-filtered copper Kα radiation (λ = 0.154178 nm). Also, powder photographs were taken with 114.6 mm radius Debye-Scherrer camera. Preliminary crystallographic investigation of single crystals (0.11 mm edge length) of the metal inclusion complex were performed with a SYNTEX P1 four-circle computer-controlled diffractometer. Molybdenum radiation Kα (λ = 0.071069 nm) was used throughout. The cell constant a_0 = 2.4664(2) nm was determined by a least-squares treatment of 15 intense reflections.

RESULTS AND DISCUSSION

The inclusion complexes $Ag_{12}A \cdot 9.5AgNO_3 \cdot 5.9H_2O$ and $Li_{12}A \cdot 9.8LiNO_3 \cdot 9.3H_2O$ can be prepared by the inclusion of molten silver nitrate and lithium nitrate in channels and cages of basic cation zeolite forms. The limiting temperatures of thermal and structural stability of the inclusion complexes were obtained from the thermoanalytical data. On Fig. 1 and Fig. 2 we can see the DTA and TGA curves for the silver nitrate inclusion complex and the lithium nitrate inclusion complex, respectively. There are a lot

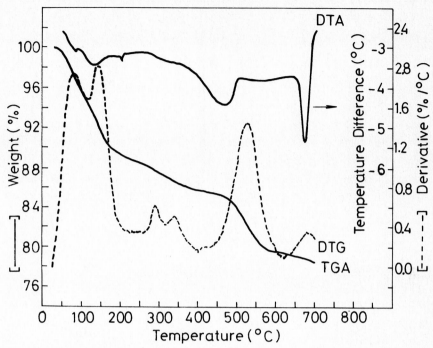

Fig. 1. DTA, TGA and DTG curves for $Ag_{12}A \cdot 9.5AgNO_3 \cdot 5.9H_2O$ inclusion complex of zeolite.

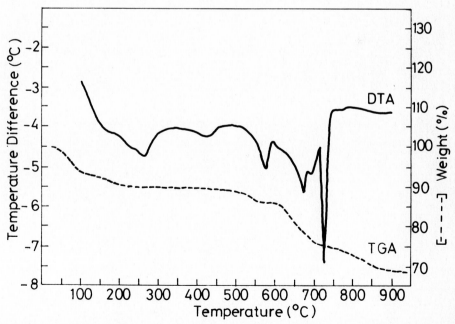

Fig. 2. DTA and TGA curves for $Li_{12}A \cdot 9.8LiNO_3 \cdot 9.3H_2O$ inclusion complex.

of thermal effects: dehydration, denitration and structural transformation. The endothermic peak of melting of silver nitrate at
the temperature of $212^{o}C$ can also be observed on the DTA curve of
the silver nitrate inclusion complex (see Fig. 1). The dehydration
takes place in the low temperature range, up to $300^{o}C$, and the denitration occurs at higher temperatures around $500^{o}C$. However, the
DTA curves of both complexes are rather different in this temperature range. The DTA curve of the $Ag_{12}A \cdot 9.5AgNO_3 \cdot 5.9H_2O$ complex
shows an endothermic peak of nitrate decomposition, while the
$Li_{12}A \cdot 9.8LiNO_3 \cdot 9.3H_2O$ complex shows three endothermic peaks. The
exact temperature of decomposition is determined by the TGA analysis using DTG curves. The temperature of denitration of the silver
nitrate inclusion complex is $520^{o}C$. The decomposition of the lithium nitrate inclusion complex is complicated and occurs in a
wide temperature range (see Fig. 2) causing the structural change
at the same time.

In the silver nitrate inclusion complex, the limiting temperatures of the included nitrate thermal stability ($520^{o}C$) and the
zeolite framework structure stability ($675^{o}C$) are rather distant,
so it is possible to make a complete inclusion complex denitration.
In the case of lithium nitrate complex, it was not possible to get
the denitrated zeolite complex, because the complete nitrate re-
moving is accompanied by the destroying of three dimensional
framework cages.

The process of denitration is followed by the changes of the
nitrate band at ca. 1400 cm^{-1} in the IR spectrum of the solid
phase (see Fig. 3). The inclusion complex shows a very strong band
at ca. 1400 cm^{-1}. From these IR spectra it was also possible to see,
via the double four ring band at 550 cm^{-1} and the pore opening
band at 380 cm^{-1}, if the three dimensional framework of cages and
channels remains unchanged. Looking at the curve 3 (see Fig. 3), we
can see that the denitration had occured at the temperature of
$520^{o}C$ in vacuum (0.13 Pa) and had lasted for 200 minutes. The nitrate bands completely disappear and the fundamental bands remain
unchanged.

On Fig. 4 we can see thermal characteristics of the metal inclusion complex. The endothermic peak at $175^{o}C$ is due to the dehydration (4.8% weight loss), and than, the sample is stable up to
$690^{o}C$, where it undergoes a structural transformation through an
endothermal effect followed by 2.5% weight loss. X-ray analysis

shows that alpha cages of 4A structure are broken.

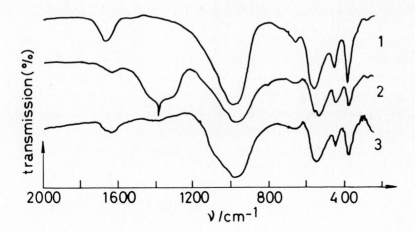

Fig. 3. Infrared spectra: (1) $Ag_{12}A \cdot nH_2O$ (silver zeolite); (2) silver nitrate inclusion complex; (3) silver inclusion complex obtained by denitration at the temperature of 520°C during 200 minutes of heating.

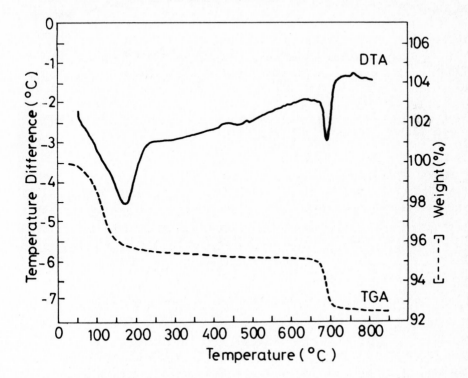

Fig. 4. DTA and TGA curves for silver inclusion complex.

X-ray studies

Powder patterns of $Ag_{12}A \cdot nH_2O$ zeolite and silver inclusion complex are given in Fig. 5. The powder data, which were measured and indexed to the cubic cell constant are given in Table 1. The powder pattern of the silver inclusion complex clearly shows that the crystal structure, formed after denitration, is stable. At the same time, by comparing with $Ag_{12}A \cdot nH_2O$ zeolite powder pattern, we can notice some differences. In our opinion the increase of d_{100} from 1.21 ($Ag_{12}A \cdot nH_2O$ zeolite) to 1.41 nm for silver inclusion complex is very important. This increase points out that the crystal lattice has been changed during the process of denitration.

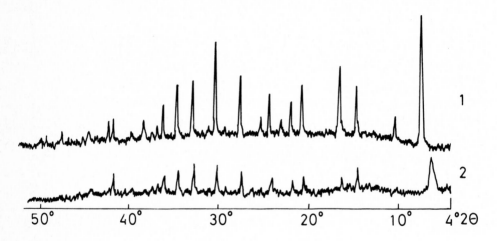

Fig. 5. X-ray powder diffraction patterns from $Ag_{12}A \cdot nH_2O$ zeolite (1) and silver inclusion complex (2).

The powder pattern of the silver inclusion complex can be indexed only to the F lattice with a_0 = 2.4664(2) nm. The reflection (111) with d = 1.41 nm for the silver inclusion complex does not present a reflection of a "superstructure" as is the case with the reflection at angle 2Θ = 21.3° for NaA zeolite (ref. 3). This cannot be explained by ordering of Al and Si in the zeolite framework alone (ref. 4).

The logical explanation is that the included Ag cations after the process of denitration are arranged in alpha and beta cages in such a way that they cause a doubling of the cubic cell constants, and a lowering of symmetry from Pm3m to Fm3c. Our preliminary structure (ref. 5) determinations confirm that and give us an indication of the presence of Ag cations in beta cages. Their arrangement is similar to the octahedral hexasilver molecule (ref. 6) found by Y. Kim and K. Seff.

TABLE 1

X-ray powder diffraction data for $Ag_{12}A \cdot nH_2O$ zeolite and silver inclusion complex.

Ag_{12}-A·nH_2O zeolite			Silver inclusion complex		
HKL	Int.	$d_{obs.}$ (nm)	HKL	Int.	$d_{obs.}$ (nm)
100	100	1.2136	111	100	1.4118
110	20	0.8585			
200	40	0.6o98	400	65	0.6108
210	60	0.5463	420	45	0.5464
220	40	0.4323	440	45	0.4322
221,300	30	0.4074	600,442	35	0.4069
310	10	0.3863			
311	35	0.3687	622	55	0.3694
222	15	0.3533	444	25	0.3529
			640	25	0.3402
321	50	0.3272	642	70	0.3269
400	5	0.3062	800	25	0.3053
322,410	80	0.2972	820,644	90	0.2965
330,411	10	0.2886	822,660	20	0.2880
331	5	0.2813			
420	50	0.2739	840	90	0.2735
			911,753	20	0.2685
332	40	0.2611	664	65	0.2608
422	30	0.2500	844	45	0.2499
430,500	10	0.2451	1000,860	25	0.2448
431,510	5	0.2405	1020,862	20	0.2405
333,511	15	0.2356			
432,520	10	0.2278	1040,864	25	0.2275
			1042	25	0.2238
440	20	0.2167	880	50	0.2162
441,522	20	0.2134	1044,882	20	0.2133
442,600	10	0.2042			
610	5	0.2019			
443,621,540	10	0.1915			
542,630	10	0.1827			
			888	20	0.1768
543,550,710	5	0.1735	10100,1086, 1420	20	0.1731

ACKNOWLEDGMENT

We thank Prof. K. Seff for supplying the crystals of NaA zeolite.

REFERENCES

1   N. Petranovic, U. Mioc, M. Susic, R. Dimitrijevic and
     I. Krstanovic, J. Chem. Soc.,Faraday Trans. I, 77 (1981),
     379-389
2   M.V. Susic, N.A. Petranovic and D.M. Mioc, J. Inorg. Nucl.
     Chem., 33 (1971), 2667
3   R.M. Barrer and W.M. Meier, Trans. Faraday Soc., 54 (1956),
     1074-1085
4   V. Gramlich and W.M. Meier, Z. f. Kristall. 133 (1981),
     134-149
5   To be published
6   Y. Kim and K. Seff, J. Am. Chem. Soc., 100 (1978), 6989-6997

QUANTUM CHEMICAL STUDY OF THE PROPERTIES OF Fe, Co, Ni AND Cr ION-
-EXCHANGED ZEOLITES

S. BERAN and P. JÍRŮ
The J. Heyrovský Institute of Physical Chemistry and Electrochem-
istry, Czechoslovak Academy of Sciences, 121 38 Prague 2
(Czechoslovakia)

ABSTRACT

The CNDO/2 method is employed in studying the properties of
$Cr(II)$, $Cr(III)$, $Cr(V)$, $Cr(VI)$, $Fe(II)$, $Fe(III)$, $Co(II)$, $Co(III)$,
$Ni(I)$, and $Ni(II)$ cations localized in the cationic positions of
zeolites modelled by clusters $T_6O_6(OH)_{12}$-cat (T=Si or Al). It is
shown that a strong donation of electron density from the zeolite
skeleton to the cations depending on the type and valency of the
cation takes place. This donation of electrons compensates the
changes in the cation properties caused by the acception or donation of electrons and influences the cation redox properties, as
well as other characteristics of clusters.

INTRODUCTION

The cations that compensate the negative charge of the zeolite
skeleton determine substantially the physical, chemical and catalytic properties of the zeolite. While nontransition metal cations
affect the zeolite properties through their electrostatic field or
through their ability to act as Lewis centres, transition metal cations may also act as redox sites in zeolites. Transition metal cations, localized in the cation position of the zeolite (as well as
randomly located cations or metal or oxide clusters) can be present
in various oxidation states in dependence on the manner of the cation introduction into zeolites and on the further zeolite treatment. The properties of the cations are also affected by the type
of cation position as well as by the fact that, in addition to coordination of cations by the skeletal oxygens, further ligands may
be bonded to them (H_2O, OH, O, etc.) (refs. 1,2).

According to the experimental data (refs. 1,2) Cr cations may be
present in the zeolite cation positions in various oxidation states: as $Cr(II)$, $Cr(III)$ and $Cr(VI)$. On the other hand, divalency
seems most probable for the Fe, Co and Ni cations located in the

cationic sites. After treating zeolite containing these cations in oxygen or hydrogen, the only trivalent cation found was Fe(III); the only lower valency cation was Ni(I) (refs. 1,2).

Along with the numerous experimental methods a further direct information (on a molecular level) on the physico-chemical properties of various forms of zeolites can be gained from quantum chemical calculation of the characteristics of model clusters. This work thus deals with theoretical study of the properties of various forms of Cr, Fe, Co and Ni faujasites.

MODEL AND METHOD

When studying the properties of the solid phase by quantum chemical methods, it is necessary to employ models of this solid phase containing a limited number of atoms - clusters. Although this procedure simplifies the conditions in a real crystal, it has been shown to be capable of providing much qualitatively reasonable information on zeolite properties (refs. 3-13).

Thus Cr, Fe, Co and Ni faujasites were modelled using the already successfully employed model clusters $T_6O_6(OH)_{12}$cat (Fig. 1)

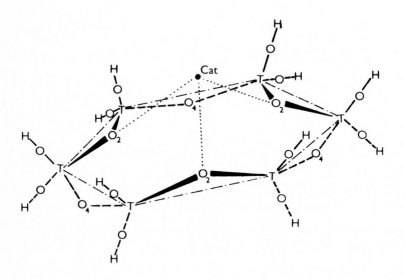

Fig. 1. Depiction of the zeolite model with indication of the individual types of atoms for the cation located in the S_{II} cationic position (symbols T stand for Si or Al atoms).

(where T is an Si or Al atom), representing six-fold windows in the zeolite directed into a large cavity or prism. The metal cations Fe(II), Fe(OH)(I), Fe(III), Co(II), Co(OH)(I), Co(III), Ni(II), Ni(OH)(I), Ni(I), Cr(III), Cr(II), Cr(V) and Cr(VI) were localized in the S_{II} of S_I' cation position and the hydroxyl group was located on the three-fold zeolite axis directed into the large cavity or cubo-octahedron. The lengths of the cation-OH and O-H bonds were 2.0×10^{-10} and 1.08×10^{-10} m, respectively (refs. 7,8). Geometric characteristics obtained from X-ray data on Ni faujasites were used for Cr, Ni and Fe zeolites and those from Co faujasites were used for Co zeolites (refs. 7,8). The geometry of the models was not adapted for the different lengths of the Al-O and Si-O bonds for the reasons given in (ref. 3). The clusters were terminated by H atoms.

Calculations on the zeolite clusters were carried out by the CNDO/2 method expanded for transition metals and parametrized by Clack et al. (ref. 14) using an s, p atomic orbital base for the Si and Al atoms (ref. 3).

RESULTS AND DISCUSSION

Model quantum chemical calculations on zeolite clusters with transition metal cations localized in the cation positions indicated that, similarly as for the cations of other metals, these cations are always bonded to three skeletal oxygen atoms (for cations in the S_{II} position, type O_2 oxygen atoms; for the S_I' position, type O_3) by electron donor-acceptor bonds. Formation of these bonds leads to transfer of electron density from the lone electron pairs of the skeletal oxygen atoms to the cation and thus to a decrease in its real charge. The magnitude of the charge transferred depends (i) on the character of the cation (on its electron acceptor ability) and, for certain valencies of the studied types of cations (e.g. divalency) increases roughly with increasing atomic number; (ii) on the formal valency of the given type of cation - increases with increasing formal valency; (iii) on the Si:Al ratio - decreases with increasing Si:Al ratio. These trends are illustrated by the values of the charge transferred, q_{trans}, given in Table 1. The amount of charge transferred between the cation and the zeolite skeleton then determines the strength of the corresponding electron donor-acceptor bond between the cation and the skeleton, as follows from comparison of the q_{trans} and $P_{cat.-O}$ values given in Table 1. Simultaneously, the amount of transferred charge is reflected

in the strengths of the Si-O and Al-O bonds of those oxygen atoms
that take part in coordination of the cation. The strength of these bonds decreases with increasing magnitude of the charge transferred, which can be explained by pointing out that the electrons
are not taken only from the lone electron pair of the oxygen, but
also from the Si-O and Al-O bonds (cf. p_{Si-O} and p_{Al-O} in Table 1).

TABLE 1

CNDO Charge Densities on Cations, q_{cat}, Wiberg Bond Orders,
p, and Energies of the Highest Occupied Molecular Orbital,
E_{HOMO}, and Lowest Unoccupied Molecular Orbital, E_{LUMO}, for
Zeolite Clusters with Cations Cr, Fe, Co and Ni Located
in the S_{II} Cationic Positions (Si:Al = 1)

Type of cation and its valency	p_{cat-O}	p_{Si-O}	p_{Al-O}	q_{cat}[a]	q_{trans}[a]	E_{HOMO}[b]	E_{LUMO}[b]
Cr(II)	0.57	0.77	0.47	1.00	1.00	- 6.4	-1.5
Cr(III)	0.64	0.67	0.44	1.20	1.80	- 8.3	-1.5
Cr(V)	0.93	0.50	0.34	2.14	2.86	-11.6	-3.9
Cr(VI)	1.14	0.41	0.30	2.54	3.46	-12.3	-6.0
Fe(II)	0.54	0.72	0.47	0.39 (0.49)	1.61 (1.51)	-13.0	-1.6
Fe(III)	0.59	0.69	0.48	1.16	1.84	-12.2	≃3.0
Co(II)	0.49	0.73	0.48	0.46 (0.57)	1.54 (1.43)	-12.9	-1.9
Co(III)	0.58	0.67	0.48	1.04	1.96	-13.3	-5.5
Ni(II)	0.57	0.69	0.59	0.31 (0.41)	1.69 (1.59)	-12.9	-1.9
Ni(I)	0.51	0.61	0.30	0.30	0.70	- 8.3	-1.7

[a] values in brackets correspond to clusters with Si:Al = ∞
[b] values correspond to the uncharged clusters

Generally, transition metal cations localized in the cationic
sites of zeolites are characterized by a relatively low charge localized on the transition metal atom, resulting from the marked
electron donor ability of the zeolite skeleton. An identical effect
is also important for the redox properties of these cations.
Changes in the cation properties (e.g. in the charge on the cation)
as a result of acceptance or donation of an electron by the cation

are compensated to a considerable degree by decreased or increased
donation of electron density from the zeolite skeleton.

Information on the ability of a molecule or cluster to donate
or accept electrons can be obtained to a first approximation from
the values of the energy of the highest occupied molecular orbital,
E_{HOMO}, or the lowest unoccupied molecular orbital, E_{LUMO}. In clusters containing transition metal cations, both these frontier orbitals are localized primarily on the cations. Consequently, these
cations represent the redox sites of the zeolite and electrons are
donated or accepted by these transition metal cations, while the
zeolite skeleton only compensates changes produced by this process
by donation of electron density. It follows from the E_{HOMO} value
that, compared with the other cations, the Cr(II), Cr(III) and
Ni(I) cations exhibit electron donor ability; i.e. exhibit a tendency to be oxidized. In contrast, the Cr(VI), Cr(V), Co(III) and
also Fe(III) cations exhibit strong electron acceptor capability;
i.e. they have a tendency to be reduced. These results thus indicate that the most probable (stable) valency for the individual types of cations should be: Cr(III) (and maybe even Cr(V)), Fe(II),
Co(II) and Ni(II). The remaining studied valencies show the above-
-mentioned tendency to be oxidized or reduced, which may be so
strong that some oxidation states of the cations were not observed
in the zeolites at all.

The well-known decrease in the stability of the zeolite skeleton
with decreasing Si:Al ratio can be explained by the different
strengths of the Si-O and Al-O bonds (ref. 12), as is apparent from
comparison of the calculated bond orders p_{Si-O} and p_{Al-O} (cf. Table 1). Metal cations localized in the cation position affect the
zeolite stability in two ways. The formation of new cation - zeolite skeleton bonds may contribute to the stabilization of the zeolite window, but, on the other hand, the Si-O and Al-O bonds involving those skeletal oxygen atoms participating in cation coordination are weakened. The strength of these new bonds is proportional
to the weakening of the Si-O and Al-O bonds, as is apparent from
comparison of the $p_{cat.-O}$ and p_{Si-O}, p_{Al-O} values given in Table 1.
With some polyvalent cations (especially Cr(VI) and Cr(V)), these
effects lead to such marked weakening of the Si-O and Al-O bonds
that these skeletal oxygen atoms are bonded to the cation more
strongly than to the Si or Al atoms of the zeolite skeleton (cf.
$p_{cat.-O}$, p_{Si-O} and p_{Al-O} in Table 1). It can thus be assumed that

transition metal cations in high oxidation states can remove oxygen atoms from the zeolite skeleton with formation of an oxide phase. This is, in fact, indicated by the decreasing stability of the zeolite skeleton with increasing oxidation state of the cation.

Cluster models of zeolites with transition metal cations bonded to a hydroxyl group represent cation models with a fourth ligand. Calculations on such models indicated that transition metal cations (in contrast to alkaline earth cations) exhibit quite high affinity for this additional ligand. Calculated cation - hydroxyl group bond orders for transition metals attain values around 0.7, i.e. values comparable with the value calculated for the $Al(OH)$ and $Al(OH)_2$ cations in zeolites (ref. 10) and substantially higher than the values found for the $Ca(OH)$ cations (ref. 9). The cation - hydroxyl group bond has the character of an electron donor-acceptor bond and its formation again leads to donation of electron density from this OH group to the cation. This is reflected in the somewhat lower value of the charge on the cation bonded to the OH group compared to the same cation without an OH group. For example, for zeolite models with $Fe(OH)(I)$, $Co(OH)(I)$ and $Ni(OH)(I)$ cations with an Si:Al ratio of 1, the charge on the transition metal equals 0.27, 0.34 and 0.18, respectively, compared with the values 0.39, 0.46 and 0.31, found for the $Fe(II)$, $Co(II)$ and $Ni(II)$ cations, respectively, It thus follows that the presence of a fourth ligand contributes to stabilization, especially of higher valencies of cations localized in zeolites. These calculations indicate that the acidity of these hydroxyl groups on transition metal cations is higher than the acidity of similar OH groups on alkaline earth cations (ref. 9), and comparable to the acidity of the hydroxyl groups of the $Al(III)$ in cationic sites (ref. 10). However, the acidity of these OH groups is substantially lower than the acidity of skeletal hydroxyl groups (ref. 11). The calculations thus indicate that the acidity of the cationic hydroxyl groups will be much smaller than that of the skeletal OH groups in agreement with experiments. This is in agreement with the experimental results concerning the strength of the acidity of various OH groups in zeolites (ref. 15).

REFERENCES
1   P.A. Jacobs, Carboniogenic Activity of Zeolites, Elsevier, New York, 1977.
2   C. Naccache and Y. Ben Taarit, Acta Physica et Chimica, 24(1) (1978) 23.

3   S. Beran and J. Dubský, J. Phys. Chem., 83 (1979) 2538.
4   W.J. Mortier and P. Geerlings, J. Phys. Chem., 84 (1980) 1982.
5   J.A. Tossel and G.V. Gibbs, J. Phys. Chem. Minerals, 2 (1977) 21.
6   E.C. Hass, P.G. Mezey and P.J. Plath, J. Mol. Structure, 77 (1981) 389.
7   S. Beran, Coll. Czech. Chem. Commun., in press.
8   S. Beran, P. Jiru and B. Wichterlová, Zeolites, in press.
9   S. Beran, J. Phys. Chem., 86 (1982) 111.
10 S. Beran, P. Jiru and B. Wichterlová, J. Phys. Chem., 85 (1981) 1951.
11 S. Beran, J. Mol. Catal., 10 (1981) 177.
12 S. Beran, Z. Phys. Chem., N.F., 123 (1980) 129.
13 S. Beran, Chem. Phys. Lett., 84 (1981) 111.
14 D.W. Clack, N.S. Hush and J.R. Yandle, J. Chem. Phys., 57 (1972) 3503.
15 J.W. Ward, ACS Monograph, No. 171 (1976) 118.

CATIONIC RHODIUM COMPLEXES AND RHODIUM METAL AGGREGATES IN ZEOLITE Y

H. VAN BRABANT, R.A. SCHOONHEYDT and J. PELGRIMS

Centrum voor Oppervlaktescheikunde en Colloïdale Scheikunde, Katholieke

Universiteit Leuven, De Croylaan 42, B-3030 Leuven (Heverlee), Belgium

ABSTRACT

A quantitative study of the CO-uptake and CO_2-production over $[Rh(NH_3)_5Cl]^{2+}$ and oxidized Rh in zeolite Y reveals the formation of $Rh(I)(CO)_2$. The conditions for its formation are delineated. The dicarbonyl is characterized by IR and reflectance spectroscopy. Oxidized Rh-zeolites contain Rh(III) partially as isolated cations, partially in the form of oxidic species. Upon reduction in H_2, it sinters at the external surface, where it is oxidized in O_2 to RhO_2.

INTRODUCTION

The chemistry of Rh in zeolites is fairly complicated. The activation of $[Rh(NH_3)_5Cl]$-Y in O_2 produces a small amount of esr-detectable Rh(II) [1]. When the O_2-pretreatment is performed at 773°K, pairs of Rh-ions are formed, thought to be $Rh^{2+}-Rh^+$ or Rh^+-Rh^0 [2]. Their maximum concentration, 20 % of the total Rh-content, was obtained on a sample loaded with 3 wt % Rh [2]. Activation in CO in the presence of H_2O occurs following the overall reaction [3] :

$$Rh(III) + 3\ CO + H_2O \longrightarrow Rh(I)(CO)_2 + CO_2 + 2\ H^+ \hspace{2cm} (1)$$

Rh(I) is a stable intermediate in the activation in vacuo of $RhCl_3$-exchanged zeolite Y especially at small loadings [4]. The final stage is Rh(0). In zeolite A, exchanged with $RhBr_3$, Rh(III) is not reduced upon activation in vacuo at 573°K [5]. It is the aim of the present study to quantify the CO- and redox-chemistry of Rh on zeolites Y and to characterize the metal phase.

EXPERIMENTAL

Samples

Rh-zeolites were prepared by exchange of NaY with $[Rh(NH_3)_5Cl]Cl_2$ at RT during 86.4 ks. The amount of $[Rh(NH_3)_5Cl]Cl_2$ (Strem Chemicals) in solution was equal to the amount calculated for the desired exchange level. The exchangeable cation contents are reported in table 1. The number after the sample symbol is

the number of Rh ions per unit cell (rounded off). The amount of protons was
calculated as the difference between the theoretical C.E.C. and the sum of Rh
and Na.

TABLE 1

Exchangeable cation contents (mmol g^{-1}) and exchange conditions of
the Rh-zeolites.

Sample	$[Rh(NH_3)_5Cl]^{2+}$	Na^+	H^+	Exchange conditions
RhY1	0.11	4.17	–	2.5 g dm^{-3}
RhY3	0.24	3.96	–	1 g dm^{-3}
RhY5	0.37	3.62	–	1 g dm^{-3}
RhY7	0.54	2.85	0.37	1 g dm^{-3}
RhY8	0.63	2.45	0.60	3.3 g dm^{-3}
RhY10	0.79	1.82	0.93	2.5 g dm^{-3}
RhY16	1.22	1.53	0.33	1 g dm^{-3} at constant ionic strength of 0.01 N with NaCl

Procedures

CO-uptake and CO_2-production were measured with the calibrated low volume
recirculation system described by Verdonck et al. [6]. The working conditions
were as described [6]. With the same apparatus CO-uptake and CO_2-production
were also measured at fixed temperatures. In every case CO_2 was trapped in a
solid CO_2 trap and determined after evacuation of CO. With a chromatographic
method [7] rates of CO-uptake and CO_2-production were determined. Integration
of the curves yielded amounts.

H_2-reduction and O_2-oxidation were performed with the same calibrated low
volume recirculation system in the same conditions as described in [6]. IR
spectra were recorded in double beam mode with a Beckman IR12 or a Perkin-
Elmer 580B. Reflectance spectra were recorded on a Cary 17 with NaY as a reference. The metal phase was characterized by X-ray line broadening after correction for zeolitic peaks by a standard peak profile analysis [8] and by H_2-
chemisorption at RT [9].

RESULTS
CO-chemistry

Fig. 1 shows - as an example - the CO-consumption and CO_2-production on
oxidized RhY5. The gaschromatographic method yields 2 rate maxima at $\leqslant 373°K$
and at 443°K, with no appreciable desorption (negative rate) below 573°K. This
is confirmed by the desorption curve of CO in the circulation system. Moreover
a significant time dependence of the CO-uptake is observed when the temperature
is kept constant at 523°K.

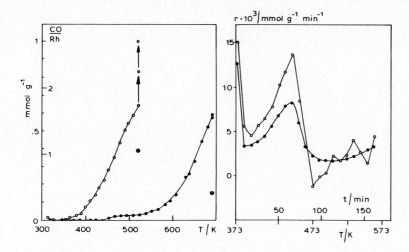

Fig. 1. Left : CO uptake (o) and CO_2 production (⊖) on RhY5 pretreated in O_2 at 623°K. Desorption of CO(●) and CO_2 (◑). The heating rate is 0.0833 K s^{-1}. Right : rate of CO consumed (o) and CO_2 produced (●) on RhY5 pretreated in O_2 at 623°K.

TABLE 2

Total CO-uptake (CO_t), CO_2-production and CO-adsorption (CO_a) of RhY5.

A. O_2-pretreatment at 623°K.

Gaschromatography			Recirculation system		
$\dfrac{CO_t}{Rh}$	$\dfrac{CO_a}{Rh}$	$\dfrac{CO_2}{Rh}$	$\dfrac{CO_t}{Rh}$	$\dfrac{CO_a}{Rh}$	$\dfrac{CO_2}{Rh}$
373°K	0.31	0.09	0.22	up to 523°K and overnight at 523°K
443°K	1.29	0.29	1.00	2.70	1.78	1.05
total	1.86	0.64	1.22	at 448°K during 5 days
				3.43	1.67	1.76

B. Samples as such.

Gaschromatography			Recirculation system		
$\dfrac{CO_t}{Rh}$	$\dfrac{CO_a}{Rh}$	$\dfrac{CO_2}{Rh}$	$\dfrac{CO_t}{Rh}$	$\dfrac{CO_a}{Rh}$	$\dfrac{CO_2}{Rh}$
403°K	1.57	1.11	0.46	isothermal steps up to 473°K
505°K	0.72	-	0.74	1.98	1.34	0.64
total	2.29	1.11	1.20	isothermal steps up to 440°K with long equilibration times at each
523°K	0.73 (desorption)			temperature
				2.69	1.73	0.96

RhY5 as such has also 2 maxima in the rates of CO consumption and CO_2 production around 403°K and 505°K. The latter is immediately followed by desorption of CO with a negative rate minimum at 523°K. This indicates that the 505°K maximum corresponds to the reduction to the metallic state. The amounts of CO consumed (CO_t), CO_2 produced and CO adsorbed ($CO_a = CO_t - CO_2$), expressed relative to the Rh content of the zeolites, are tabulated in table 2 for several experiments. Two remarks can be made about table 2.

(i) The rate maxima of CO-uptake and CO_2-production for RhY as such and oxidized Rh-zeolites not only occur at different temperatures but also the amounts at each maximum are different. This indicates different reduction phenomena.

(ii) In no case is the stoichiometry implied by reaction (1) obtained. This stoichiometry is approached only after relatively long reaction times (overnight at 523°K for oxidized RhY5; isothermal steps up to 440°K with long equilibration times at each temperature for RhY5 as such). Very long reaction times give more CO-consumption and CO_2-production but less adsorbed CO than required by (1). With short time experiments the ratios CO_t/Rh, CO_a/Rh and CO_2/Rh are smaller than required by reaction (1).

On oxidized samples we have obtained the same IR spectra upon adsorption of CO as published by Primet et al. [3]. Thus, 2 dicarbonyls $Rh(I)(CO)_2$ are formed with slightly different frequencies, 2025-2100 cm^{-1} (LF) and 2045-2115 cm^{-1} (HF), the latter being the predominant species. With thick films weak bands at 2140 cm^{-1} and 2170 cm^{-1}, ascribed to Rh(III)-CO can also be observed. However, we have never obtained evidence of consumption of OH groups or H_2O according to reaction (1).

When CO is adsorbed on the samples as such, the LF carbonyl starts to appear after evacuation at 343°K (fig. 2). This is also the start of the decomposition of $[Rh(NH_3)_5Cl]^{2+}$ as evidenced by the formation of NH_4^+ (1440 cm^{-1}) and the shift of the 1340 cm^{-1} band of coordinated NH_3 [10] to 1310 cm^{-1}. At the same time Rh(III)-CO is formed, absorbing at 2165 cm^{-1} [3]. This band shifts to 2175 cm^{-1} at higher temperatures and disappears above 423°K. The 2 bands at 2230 cm^{-1} and at 2260 cm^{-1} are believed to be due to isocyanate species [11]. We do not know if they are surface-bonded or Rh(III)-NCO species [12]. The HF carbonyls appear after evacuation at 373°K and become the dominant species at higher temperatures. Both carbonyls are destroyed by evacuation at 523°K. The sharp band at 1405 cm^{-1} is due to NH_4Cl condensed on the cell windows. It is never found during an oxidative pretreatment. Hydroxyl groups, absorbing at 3750 cm^{-1}, 3660 cm^{-1} and 3550 cm^{-1} can only be clearly detected after high temperature evacuation. They originate from proton exchange during sample preparation and from decomposition of NH_4^+.

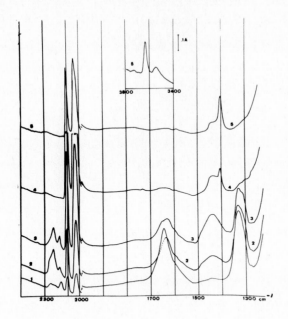

Fig. 2. IR spectra of CO on RhY7. The spectra are obtained at RT after adsorption (3.6 ks) and evacuation (3.6 ks) at the indicated temperatures : 1, 343°K; 2, 373°K; 3, 416°K; 4, 473°K; 5, 523°K.

When the samples as such are heated in a $CO:H_2O$ (1:1) mixture up to 473°K, only the LF dicarbonyl is formed. There is no indication in the IR spectra of isocyanate and Rh(III)-CO species. In the reflectance spectra (fig. 3) 2 bands at 320 nm and 255 nm emerge, stable up to 423°K. The band positions agree respectively with the $Rh4dz^2 \longrightarrow CO\ 2\pi^{::}$ and $Rh4dxz,4dyz \longrightarrow CO\ 2\pi^{::}$ transitions of $[Rh(CO)_2Cl]_2$ [13].

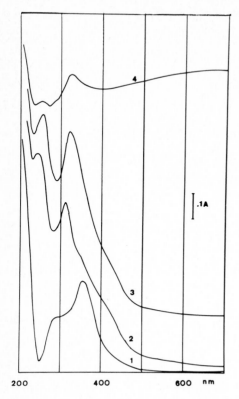

Fig. 3. Reflectance spectra of RhY5 after treatment during 7.2 ks in a CO/H$_2$O flow at the indicated temperature. The gasflow is 3 vol. % CO in He bubbling through a H$_2$O saturator at the desired temperature for a 1:1 CO/H$_2$O ratio. 1, 295°K; 2, 403°K; 3, 423°K; 4, 448°K.

Redox chemistry

In table 3 we report the quantitative H$_2$ and O$_2$-uptakes on RhY8. After pretreatment in O$_2$ at 773°K the H$_2$-uptake between RT and 773°K agrees quantitatively with the reduction of Rh(III). Two reactions are possible :

$$2 \, Rh(III) + 3H_2 \longrightarrow 2 \, Rh(0) + 6H^+ \tag{2}$$

$$Rh_2O_3 + 3H_2 \longrightarrow 2 \, Rh(0) + 3H_2O \tag{3}$$

H$_2$O-production is always observed but cannot be quantitatively measured due to condensation. Therefore, the relative importance of reactions (2) and (3) cannot be assessed. However, in table 4 it is shown that the OH-band intensities of RhY10 reversibly increase after reduction and decrease after oxidation. If the total band area after oxidation corresponds to 0.93 mmol H$^+$ g^{-1} (table 1), the

TABLE 3

H_2- and O_2-uptakes during redox cycles of RhY8.

Cycle	H_2 mmol g^{-1}	O_2 mmol g^{-1}	$\dfrac{O_2}{Rh}$	$\dfrac{(O_2)_n}{(H_2)_{n+1}}$	$(\dfrac{O_2}{H_2})_n$
A. Pretreated in O_2 at 773°K					
1	0.93	0.69	1.10	–	0.74
2	1.45	0.68	1.08	0.48	0.47
3	1.48	0.69	1.10	0.46	0.47
4	1.47	0.65	1.03	0.47	0.44
5	1.49	0.64	1.02	0.44	0.43
B. Pretreated in O_2 at 623°K					
1	0.71	0.31	0.49	–	0.44
2	0.80	0.33	0.52	0.39	0.41
3	0.89	0.32	0.51	0.37	0.36
C. Pretreated in vacuo at 773°K					
1	–	0.61	0.97	–	–
2	1.43	–	–	–	–
3	1.39	–	–	–	–

TABLE 4

Variation of OH band intensities upon reduction and
oxidation of RhY10.

Treatment	3650 cm^{-1} (mm^2)	3550 cm^{-1} (mm^2)	Total (mm^2)
O_2 at 733°K	523	586	1109
H_2 at 733°K	847	837	1684
O_2 at 733°K	513	544	1057
H_2 at 733°K	764	863	1627

OH band increase after reduction corresponds to 0.49 mmol g^{-1} or 0.16 mmol Rh
are reduced according to reaction (2).

Once reduced, reaction (4) and (5) occur quantitatively :

$$Rh(O) + O_2 \longrightarrow RhO_2 \tag{4}$$

$$RhO_2 + 2H_2 \longrightarrow Rh(O) + 2H_2O \tag{5}$$

Whatever the redox cycle, reduction is complete below 400°K, oxidation is complete at 773°K. When the sample is pretreated at 623°K in O_2 and the redox cycles are performed between RT and 623°K, the same reactions (2)-(5) occur but the extent of the reactions is only about 50 % of the total Rh-content. A pretreatment in vacuo at 773°K reduces all the Rh(III) to the metallic state and there is no measurable H_2 uptake in the first cycle. The subsequent O_2-interaction follows reaction (4).

When RhY8 and RhY10 are reduced in H_2 at 673°K the Rh-metal peak appears in the X-ray diffractograms at $2\theta = 41°$. The average metal particle size calculated from the width of peak is 4.6 nm and 8.2 nm for RhY8 and RhY10 respectively. For RhY1, similarly reduced, no X-ray peak of metallic Rh is observed. The H_2 chemisorption technique gives a H:Rh ratio of 0.12.

DISCUSSION

The uptake of CO, the production of CO_2 and the IR-evidence for a dicarbonyl confirm reaction (1), first proposed by Primet et al. [3]. On the basis of our data a number of supplementary remarks can be made.

(i) The reaction also occurs on samples as such. In that case, the dicarbonyls are formed in one step as soon as the $[Rh(NH_3)_5Cl]^{2+}$ starts to decompose i.e. above 323°K in vacuo. The maximum rate of formation is at 403°K. Side reactions are the formation of isocyanate and of Rh(III)-CO.

(ii) On oxidized zeolites Rh(III) is reduced to $Rh(I)(CO)_2$ in 2 steps below 523°K. This may correspond to the reduction of a small amount of Rh(III) ($T_R \leqslant 373°K$) and of Rh_2O_3 ($T_R = 443°K$). It may also be that only one type of Rh is present, mainly localized in the sodalite cages ($T_R = 443°K$) with a small amount in the supercages ($T_R \leqslant 373°K$). H_2-reduction of oxidized samples is in agreement with the first hypothesis. Whether after an oxidative treatment Rh(III) is present as Rh_2O_3 or as a monomeric or dimeric Rh(III)-oxide species is still under investigation. In any case its presence explains why no OH or H_2O molecules are consumed upon reduction with CO. The reaction with CO can formally be written as :

$$Rh_2O_3 + 7 \ CO \longrightarrow 2 \ Rh(I) \ (CO)_2 + 3 \ CO_2 \qquad (6)$$

(iii) It is not clear whether the occurrence of LF and HF dicarbonyls is due to different ligands (number, kind) in the coordination sphere or to different siting. In this respect it is remarkable that the IR frequencies of the dicarbonyls are independent of the pretreatment. In any case between 403°K and 443°K and in the presence of water the reflectance spectra give evidence for $(OC)_2-Rh \underset{Cl}{\overset{Cl}{\lessgtr}} Rh-(CO)_2$, with only the LF carbonyl bands pair in IR. The HF dicarbonyl may then be due to $Rh(I)(CO)_2$ moieties without Cl^-.

Reduction with H_2 gives metallic Rh on the external surface. There is no evidence for a bidisperse metal particles distribution. This is a quite different behaviour from Ru [6].

ACKNOWLEDGMENT

Acknowledgment is made to the donors of the Petroleum Research Fund, administered by the American Chemical Society, for support of this research. R.A.S. acknowledges his position as Senior Research Associate of the National Fund of Scientific Research (Belgium). The authors thank Dr. W.J. Mortier for his aid with the peak profile analysis. The continuous interest of Prof. J.B. Uytterhoeven was a great stimulus.

REFERENCES

1 C. Naccache, Y. Ben Taarit and M. Boudart, A.C.S. Symp. Ser., 40 (1972) 156-165.
2 V.D. Atanasova, V.A. Shvets and V.B. Kazanski, Kin. Katal., 18 (1977) 753-757.
3 M. Primet, J.C. Vedrine and C. Naccache, J. Molecular Catal., 4 (1978) 411-421.
4 Y. Okamoto, N. Ishida, T. Imanaka and S. Teranishi, J. Catal., 58 (1979) 82-94.
5 M. Kuznicki and E.M. Eyring, J. Catal., 65 (1980) 227-230.
6 J.J. Verdonck, P.A. Jacobs, M. Genet and G. Poncelet, J. Chem. Soc. Faraday I, 76 (1980) 403-416.
7 J.J. Verdonck, P.A. Jacobs and R.A. Schoonheydt, submitted to J. Phys. Chem.
8 W.J. Mortier, N.B.S. Special Publication 567. Proc. Symp. Accuracy in Powder Diffraction, Gainesburg (1979) 315-324.
9 J.R. Anderson, Structure of Metallic Catalysts, Academic Press, London (1975) p. 295 and 365.
10 Yu.Ya. Kharitonov, N.A. Knyazeva, G.Ya. Mazo, I.B. Baranovskii and N.N. Generalova, Russian J. Inorg. Chem., 16 (1971) 1050-1054.
11 K. Nakamoto, Infrared and Raman Spectra of Inorganic and Coordination Compounds, 3rd ed., J. Wiley & Sons, New York (1978) p. 276.
12 J. Raskó and F. Solymosi, J. Catal., 71 (1981) 219-222.
13 J.F. Nixon, R.J. Suffolk, M.J. Taylor, J.G. Norman, Jr., D.E. Hoskins and D.J. Gmur, Inorg. Chem., 19 (1980) 810-813.

MECHANISMS OF FORMATION AND STABILIZATION OF METALS IN THE PORE STRUCTURE OF
ZEOLITES

P.A. JACOBS

Centrum voor Oppervlaktescheikunde en Colloïdale Scheikunde, Katholieke
Universiteit Leuven, De Croylaan 42, B-3030 Leuven (Heverlee) (Belgium)

CONTENT

1. INTRODUCTION

With the advent of the pioneering studies of Rabo et al. (ref. 1) and Dalla Betta and Boudart (ref. 2), it was realized at an early stage that zeolites could be promising supports for the preparation of very small metal clusters. As appears from the reviews of Jacobs (ref. 3), Uytterhoeven (ref. 4), Minachev et al. (ref. 5) and Gallezot (ref. 6), an extensive body of literature exists treating the system zeolite-metal particles.

From early work (ref. 3-5) it strikes the reader that :

. the major effort has been on faujasite-type zeolites (X,Y) and A. Only a minor number of studies exist treating other structures (as mordenite (MOR) and zeolite Linde L).

. the main reducing agent used has been molecular hydrogen.

. the reduction started mostly from the zeolite exchanged with the hydrated transition ions, or in case of the group VIII elements from the ammine complexes.

. reduction of almost all elements with an electrochemical potential equal to or more positive than the one of Fe(II) has been tried.

. the equations of the reduction with molecular hydrogen of stoichiometrically exchanged zeolites have been determined and are reasonably-well understood.

. detailed kinetic reduction mechanisms have only been advanced in the case of Ag(I), Cu(II) and Ni(II).

Although the early experiments seemed very promising for industrial application of these systems, it is surprising that until now only processes exist which use exclusively Pt as active material and always in combination with an acid support (ref. 7) : hydrocracking, shape selective cracking and isomerization are typical examples. In such processes the only catalytic function of the Pt-phase is to carry out - in terms of the classical bifunctional mechanism for catalysis (ref. 28) - a structure-insensitive reaction : i.e. the dehydrogenation-hydrogenation of saturated and unsaturated hydrocarbons, respectively. In such a system only sufficient Pt-surface area is needed and the average particle size and the size distribution is less important.

In view of all this, it is the aim of the present review, to make the state of the art, with respect to the mechanism of preparation of a metal phase in zeolites and to the regenerability of the system. Only if this is sufficiently-well understood, there is hope to exploit catalytically the "peculiar" properties of metals in zeolites - if they exist - and to develop industrial applications.

2. METHODS FOR THE PREPARATION OF A METAL PHASE IN ZEOLITES USING STOICHIO-METRICALLY-EXCHANGED ZEOLITES

2.a. Reduction with molecular hydrogen

The experimental conditions to be used in order to prepare stoichiometrical-ly-exchanged transition-ion zeolites in aqueous conditions have been described in detail for CuY and NiY-zeolites (ref. 9). Only very diluted exchange conditions ($\leqslant 0.02$ M) seem to be effective to reduce the formation of partially hydrolyzed species in the faujasite supercage. It has been shown recently (ref. 10) that the precipitation of a hydroxide-type phase at the outer sur-face of the zeolites inevitably occurs unless special precautions are taken. The ammine complexes of the group VIII ions show enough resistance against hydrolysis in the 5-7 pH range (ref. 11) and can be loaded on the zeolite by cation exchange in the pH-range mentioned.

The difference in reducibility of isolated transition ions (in aqueous or zeolitic phase) and as oxide (supported or not) can be understood as follows. For Me-oxides (MeO) and Me-salts in aqueous solution, the reduction occurs according to the following equations (ref. 12) :

$$MeO_{x(s)} + x\ H_{2(g)} \rightleftharpoons Me_{(s)} + x\ H_2O_{(g)}$$

$$MeCl_{x(s)} + \frac{x}{2}\ H_{2(g)} \rightleftharpoons Me_{(s)} + x\ HCl_{(g)}$$

where the subscripts s and g denote the solid and gas phase, respectively. The reducibility will be determined by the P_{HCl}/P_{H_2} and P_{H_2O}/P_{H_2} ratios for the metal oxides and salts, respectively. These ratios are plotted against each other in Fig. 1. It is clear that the reducibility of all the oxides - except Ag - will be higher than of the corresponding ions in aqueous solutions. Reduc-tion of transition ions in zeolites with molecular hydrogen is similar to the phenomena in the aqueous phase, as can be derived from the well-documented reduction stoichiometry (ref. 3-4) :

$$Me^{+n}\ Zeol.^{-n} + \frac{n}{2}\ H_2 \longrightarrow Me^o + n\ H^+\ Zeol.^{-n}$$

It therefore may be expected that the reducibility of MeO and of partially hydrolyzed species in zeolites will show enhanced values compared to that of isolated transition ions.

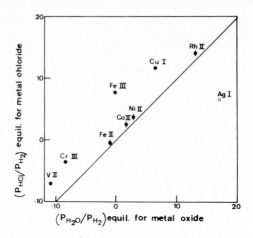

Fig. 1. Comparison of the reducibilities of MeO_x and $MeCl_x$ in water, based on the values of the P_{HCl}/P_{H_2} and P_{H_2O}/P_{H_2} of the reduction equations (values are from ref. 12).

2.b. Reduction with atomic hydrogen

Attempts to reduce NiX zeolites with atomic hydrogen have been reported recently (ref. 13,14). It seems that the reducibility of the Ni ions is enhanced (Table 1) considerably and a more complete reduction can be obtained at lower temperature. This results in the formation of a more dispersed metal

TABLE 1

Reducibility of $Ni_{31}X$ with molecular and atomic hydrogen.

reducing agent	reducing temperature/K	$\alpha^{(1)}$	particle diameter nm			ref.
H_2	573	0.19	1.5	6.0	10.0	15,16
$H^{\cdot}$	273	0.60	1.0	–	–	14

(1) degree of reduction.

phase. This increased reducibility can be expected on thermodynamic grounds, since by using hydrogen atoms rather than molecules, the standard free energy change for the reduction reaction will become more negative by some 364-440 kJ/mole of hydrogen involved (ref. 17).

It is therefore anticipated that reduction of Fe(II)Y zeolites eventually will become possible by using hydrogen atoms instead of molecules.

In situ production of "activated hydrogen" species seems also possible over pre-reduced Pt or Pd clusters in the zeolites (ref. 15,18,19). This also

results, for straightforward thermodynamic reasons, in an increased reducibility of other transition ions (e.g. Ni) and therefore will give a better final
metal dispersion. The question arises whether in these bi-ionic systems bimetallic clusters can be formed in the supercages. Results on alloy formation in
Cu-Ni zeolites show only evidence for the existance of external alloy particles (ref. 20,21) which at least catalytically behave in a way similar to
supported alloys. Ru-Ni and Ru-Cu clusters (ref. 22,23) also seem to be located
exclusively at the external surface. This suggests that metal segregation has
occurred for particles with dimensions smaller than the zeolite supercage.

2.c. Reduction with ammonia

The reduction of NiNaY with ammonia gas compared to molecular hydrogen is
given in Table 2.

TABLE 2

Reduction of NiNaY at 723 K with H_2 and NH_3.

reducing agent	α	$(Ni^o)_i$ [1]	$(Ni^o)_e$ [2]
NH_3	0.35	43	57
H_2	0.13	69	31

after ref. 24.
(1) nickel clusters inside the supercages; (2) nickel crystallites at the
external surface of the zeolite.

The increased reducibility of Ni(II) in presence of NH_3 can be attributed to a
cation location effect. Indeed, it is straightforward to assume that hidden
cations come to more accessible locations. In reduction conditions, catalytic
decomposition of NH_3 occurs at a fast enough rate to provide molecular hydrogen
or some "activated hydrogen" needed for further reduction to proceed. It also
results that the sintering rate in ammonia is increased of $(Ni^o)_i$ to $(Ni^o)_e$.
This perhaps indicates that the sintering species is a positively charged Ni-
ammine complex.

2.d. Reduction with metal vapors

The stoichiometry of such reduction should be as follows :

$$Me(II) \; Zeol.^{-2} + 2 \; Na \longrightarrow Me(0) + Na_2^+ \; Zeol.^{-2}$$

The feasibility of the reduction has been demonstrated for Ni(II) (ref. 25) and
Fe(II) (ref. 26). In case of Fe(II)Y (ref. 26), $(Fe^o)_i$ clusters are formed with

a narrow size distribution centered around 1.3 nm, although there is evidence
at the same time for external iron crystallites. It seems therefore that
irrespective of the reducing agent and the nature of the metal, there occurs
always sintering from the supercages to the external zeolite. This sintering
process may be substantially reduced at very low reduction temperatures (e.g.
reduction with Na in liquid ammonia (ref. 27)).

3. REDUCTION OF A-STOICHIOMETRICALLY EXCHANGED TRANSITION ION ZEOLITES WITH MOLECULAR HYDROGEN

For reasons explained earlier partially hydrolyzed species of transition
ions should show enhanced reducibility. This in its turn gives greater reducibility at a given temperature, or requires lower reduction temperatures to reach
the same degree of reduction. Lower reduction temperatures then give a better
dispersion and a lower degree of sintering to the external zeolite surface.
All this is illustrated for NiY (Table 3). Three samples are considered : a
stoichiometrically exchanged zeolite (NiY), a NiY calcined "deep-bed" in presence of steam to provoke cation hydrolysis NiY-DB and a NiY sample in which
$Ni(OH)_2$ was precipitated with in situ Na_2CO_3 treatment of the parent sample
(NiY").

TABLE 3
Influence of pretreatment of NiY-70 on its reducibility with molecular hydrogen.

sample	α	$(Ni^o)_i$ %	d_i nm	$(Ni^o)_e$ %	d_e nm
NiY-673[a]	0.48	45.9	1.6	2.1	16.2
NiY-DB-673[a]	0.45	35.0	0.7	10.0	15.0
NiY"-673	1.00	60.0	1.1	40.0	16.0

a, from ref. 28.

It seems that by acting on the chemical nature of the transition ion, lower
reduction temperatures and therefore better dispersions and lower degrees of
sintering to the outer surface of the zeolite can be obtained. This principle is
of general applicability since similar results are obtained for Fe(II)Y (ref.
29).

The influence of residual water during reduction of $Ru(NH_3)_6^{3+}$ in Y zeolite
determines the extent to which hydrolysis of the complex occurs (ref. 35).
Fig. 2 shows that only thoroughly degassed samples should be contacted with
hydrogen. Since during thermal decomposition of these complexes reduction already occurs to a major extent, it is clear that the sintering rate of Ru^o is

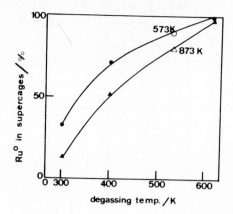

Fig. 2. Ru^o dispersion against the degassing temperature prior to reduction of the Ru(III)-hexammine Y zeolite. Two reduction temperatures are shown (573 K and 873 K). After ref. 35.

enhanced in the presence of water, which may suggest that Ru sinters as : Ru^{x+}(OH)$_y$, i.e. a charged partially hydrolyzed species.

4. PREPARATION OF METAL PHASES ASSOCIATED WITH ZEOLITES THROUGH DECOMPOSITION OF SORBED CARBONYL COMPLEXES

Adsorption of volatile metal carbonyls can be used to make a deposit of metals which otherwise are difficult to reduce. Mo, Re, Ru, Ni and Fe-carbonyls have been adsorbed (ref. 30-34). If the support exhibits residual acidity, part of the carbonyl is oxidized upon thermal activation and some metal is therefore lost irreversibly. For complexes which can be thermally decomposed at low temperature or after decomposition in UV radiation, an improved dispersion is obtained (ref. 33). In order to obtain a monodispersed metal phase in the faujasite supercages, the thermally activated diffusion processes which occur before the first ligands disappear must be suppressed (ref. 34). This can be done by heating quickly in the presence of inert gases or by performing the decomposition at low temperatures using UV radiation.

5. FACTORS WHICH INFLUENCE THE REDUCIBILITY OF TRANSITION IONS
Nature of the cation

For a given zeolite, the reduction of a transition ion, according to Uytterhoeven (ref. 4) depends on the redox potential of the ion and the following series of elements should exist, indicating increased reducibility :

Zn(II) < Fe(II) < Co(II) < Ni(II) < Cu(II) < Ag(I)

Klier et al. (ref. 36) have shown that for a series of bivalent ions in zeolites the reducibility depends on the sum of the first and second ionization potential and the stabilization energy at a given site. For Me^{+2} ions coordinated in six-membered rings, the reducibility series should be as follows :

Fe(II) < Co(II) < Ni(II) < Zn(II) < Cu(II)

If only reduction to the respective metal atoms is considered, instead of reduction to bulk metal, the Cu and Zn interchange their positions.

Although the second explanation from the theoretical point of view is quite attractive, it is difficult to explain why Zn(II) in stoichiometrically ex-changed faujasites is even much harder to reduce than Co(II) (ref. 4).

Cation location

In their study on the reduction of AgY, Beyer et al. (ref. 36) found that the last 16 Ag^+ ions were most difficult to reduce. Since the apparent activation energy for the reduction of these ions was similar to the one for cation mobility, it was concluded that the hidden sites in faujasites are difficult to reduce. In the mean time (ref. 37,38), it is shown that the resistance to reduction is the result of the formation of charged clusters, stabilized in the sodalite cages.

There seems to exist increasing agreement now that Ni(II) will be reduced preferentially in the hexagonal prisms (ref. 24,39-41). The argumentation is derived from cation-location studies on partially reduced samples. It should however be kept in mind that these structures are determined at ambiant tempera-ture, and that the zeolite-cation association is not a static system. Therefore, after preferential reduction of ions in the zeolite supercages, a rearrangement of the cation population over the different sites will certainly occur and also might result in a preferential depopulation of site I by the reducible cations. Anyway, there is ample evidence for the effect of a second cation on the reducibility of another one (ref. 38-40). High-temperature X-ray studies are urgently needed to enravel and understand all possible influences.

Chemical composition of the zeolite

It is a well-known phenomenon that the reducibility of transition ions increases in the direction (ref. 3,4,42,43)

Y < X < A zeolite

Until now, no adequate explanation has been formulated for this effect.

6. ATTEMPT TO RATIONALIZE THE REDUCIBILITY OF TRANSITION IONS IN ZEOLITES

If the system is considered in a dynamic way, it is straightforward to assume a continuous rearrangement of the cations over the sites which drives it to the thermodynamically most stable situation. It may be true that as a result of the preferential fixation of cations in certain sites or of the formation of charged clusters in sodalite cages (ref. 38), the reducibility of an ion changes with the degree of reduction. In the present treatment, only initial reduction is considered of a cation located in a zeolite supercage. The chemical composition of the surrounding zeolite will determine the electrostatic interaction of the ion with the framework and therefore its reducibility. This effect of chemical composition can be quantified using the Sanderson electro-negativity model (ref. 44-46).

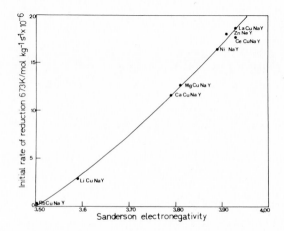

Fig. 3. Initial rate of reduction by hydrogen of transition ion zeolites with variation of their overall electronegativity, calculated according to Sanderson.

Fig. 3 shows the proportional change of zeolite electronegativity and initial rate of reduction of the transition metal ion, which is expected when the above reasoning is true. In principle, this accounts also for the variation of redu-cibility with the Si/Al ratio of the zeolite, although the data to proof it are lacking at this moment.

7. SOME TYPICAL ASPECTS OF THE REDUCTION MECHANISM OF TRANSITION IONS IN
   ZEOLITES

Hydrogen-uptake curves

   Hydrogen-uptake rates by transition ions in zeolites in all (ref. 18,47,48)
- except one (ref. 36) - cases fit to a diffusion limited model. These models
are very similar to those proposed for metal oxides (ref. 49). This agreement
may be only formal as appears from the data of Fig. 4.

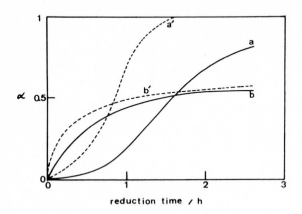

Fig. 4. Rates of hydrogen uptake of NiO at 437 K (a) and of NiY at 773 K (b) and
doped with 0.5 % wt. of Pt (a', NiO; b', NiY).

   The H_2 uptake curves are compared for bulk NiO and NiY. For the former
material, a characteristic induction period is followed by a steady progression
of the formed reaction interface. The induction period is shortened when
hydrogen activation centers (Pt) are added to the system. The behavior of the
zeolite-Ni(II) system never shows an induction period. This can be explained if
the system behaves in a dynamic way, i.e. if progression of the reduction causes
a total redistribution of the cation over the different sites - in a random way
or in a directed way when strong competition for sites exists. The slow step in
this process initially is either the hydrogen activation at residual impurities
(ref. 50) or the cation migration. As the reduction progresses a third rate-
determining component may be added to the system : the complete reduction and
sintering of charged clusters. This step can be isolated for AgY (ref. 36),
while for ions in systems with lower reducibility, they would superimpose and
show an overall progression of the reduction as shown in Fig. 4.

   An alternative explanation for the observed uptake curves would be that due
to the exothermicity of the reduction reaction and the relatively large beds

of zeolite used, significant hydrogen concentration gradients exist, resulting
in non-homogeneous reduction (ref. 51).

Decomposition of group VIII cation-ammine complexes

The group VIII cations have to be exchanged in zeolites as their ammine
complexes. It is a general observation that improved dispersion is obtained by
precalcination (ref. 6, 52-54). There exists an optimum calcination temperature
which was explained as follows (ref. 54) : this "treshold" temperature is the
one below which the decomposition of the ammine complex is too slow to occur
within a reasonable calcination time. Since this represents a kinetic effect,
depending on the temperature, the partial pressure of oxygen, the calcination
time, hydrogen attack will be either on $Pt^{2+}(NH_3)_y$ species (ref. 54) or on Pt^{2+}
ions (ref. 6,53). It explains that slight changes in the activation procedures
will affect drastically the resulting Pt dispersion.

Sintering of metals in zeolites

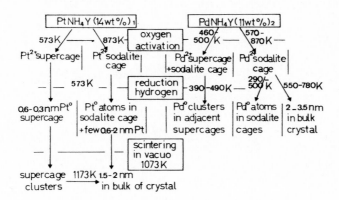

Fig. 5. Schematic representation of Pt and Pd in Y zeolites after different
reducing treatments (adapted from refs. 6 and 55).

The changes of Pt and Pd dispersions are shown in Fig. 5. It seems that ions
are reduced to Me^o atoms in the sodalite cages, which gradually sinter to supercage clusters. This sintering mechanism probably involves migration of atoms
through the six-membered windows. The particles agglomerate probably in neigh-
boring supercages (ref. 55) and in some cases can even show a grape-type
cluster (ref. 55,56). Further agglomeration of this cluster would explain that
metal particles up to 10.0 nm in the bulk of a zeolite crystal can be found
(ref. 56). This is schematically shown in Fig. 6. In agreement with the above

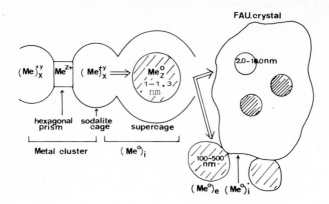

Fig. 6. Schematic representation and denotation of possible metal species in FAU-zeolites.

considerations, it has been shown that the overall sintering rate of Pt in zeolite Y in H_2 atmosphere at low temperatures (773-873 K) seems to be diffusion controlled, while at higher temperatures (973-1073 K) it is sintering-controlled (ref. 57).

The reduction of cubooctahedrally-stabilized silver clusters in A zeolite and the formation of silver particles are consecutive phenomena (ref. 58). These rate data when plotted against $t^{1/2}$ (t = time) show fairly straight lines, indicative of a diffusion phenomenon (Fig. 7). The diffusion of Ag through the six-ring window, should determine the rate. The chemical nature of the diffusing species remains to be determined.

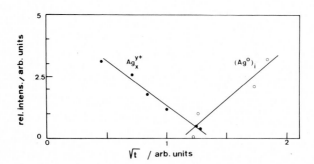

Fig. 7. Relative intensity data of Ag_x^{y+} and $(Ag^o)_i$ clusters in A zeolite against $t^{1/2}$. Original data are taken from ref. 58.

8. ATTEMPTS TO STABILIZE METAL PHASES

It was already suggested that a metal phase can be kept in the zeolite pores, provided a reduction in mild conditions would be possible. It seems now that this distribution can be preserved at higher temperatures ($\sim$ 523 K for NiCaA) when $Cr(CO)_6$ is added before reduction (ref. 59). These data do not allow to judge whether this is only the result of partial pore mouth blocking.

9. REDISPERSION OF A METAL PHASE IN ZEOLITES

Taking into account the literature data mentioned, this behavior is schematically presented in Fig. 8. Each metal particle in FAU-type zeolites

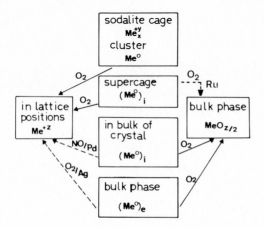

Fig. 8. Schematical representation of the reoxidation of metals in zeolites : ────── : general pathway; ---- : specific reaction for a particular element.

can be reoxidized to the initial state (charge compensating ions in lattice positions) as long as its dimensions do not exceed those of the supercage. Otherwise a bulk metal-oxide phase is formed outside the zeolite crystals. Ag^o can be completely oxidized since the oxide is very unstable. Ru^o on the other hand is always oxidized to the very stable RuO_2, irrespective of its position (ref. 35). Pd clusters in the bulk of the crystal can be reoxidized to Pd^{2+} ions in the lattice by NO treatment (ref. 60). It remains to be proven whether this redissolution is unique for Pd and eventually why.

ACKNOWLEDGMENTS

The experimental help of M. Tielen is highly appreciated. The author also gratefully acknowledges a permanent research position as "Onderzoeksleider" from the Belgian Science Foundation (N.F.W.O.-F.N.R.S.).

84

REFERENCES

1   J.A. Rabo, V. Schomaker and P.E. Pickert, Proceedings 3rd Int. Congr. Catal.,
    2, North Holland, Amsterdam, 1965, p. 1264.
2   R.A. Dalla Betta and M. Boudart, Proceedings 5th Int. Congr. Catal., 2,
    North Holland, Amsterdam, 1973, p. 1329.
3   P.A. Jacobs, "Carboniogenic Activity of Zeolites", Elsevier, Amsterdam,
    1977, p. 183.
4   J.B. Uytterhoeven, Acta Phys. Chem., 24 (1978) 53.
5   Kh.M. Minachev and Ya.I. Isakov, "Zeolite Chemistry and Catalysis", J.A.
    Rabo, ed., A.C.S., Washington, 1976, p. 552.
6   P. Gallezot, Catal. Rev. Sc. & Eng., 20 (1979) 121.
7   H. Heinemann, Catal. Rev., Sc. & Eng., 23 (1981) 315.
8   H.L. Conradt and W.C. Garwood, Ind. Eng. Chem. Proc. Des. Dev., 3 (1964) 38.
9   R.A. Schoonheydt, L.J. Vandamme, P.A. Jacobs and J.B. Uytterhoeven, J.
    Catal., 43 (1976) 292.
10 K.G. Ione, V.N. Romannikov, A.A. Davydov and L.B. Orlova, J. Catal., 57
    (1979) 126.
11 J.R. Anderson, "Structure of metallic Catalysts", Academic Press, London,
    New York, San Francisco, 1975, p. 198.
12 ibidem, p. 166-167.
13 M. Che, M. Richard and D. Olivier, J.C.S. Faraday I, 76 (1980) 1526.
14 D. Olivier, M. Richard, L. Bonneviot and M. Che, "Growth and Properties of
    Metal Clusters", J. Bourdon, ed., Elsevier, Amsterdam, Oxford, New York,
    1980, p. 165.
15 M. Briend Fauré, M.F. Guilleux, J. Jeanjean, D. Delafosse, G.D. Mariadassou
    and M. Bureau-Fardy, Acta Phys. Chem., 24 (1978) 99.
16 D. Delafosse, "Catalysis by Zeolites", B. Imelik et al., eds., Elsevier,
    1980, p. 235.
17 Y.Y. Huang and J.R. Anderson, J. Catalysis, 40 (1975) 143.
18 M.-F. Guilleux, M. Kermarec and D. Delafosse, J.C.S. Chem. Comm., (1977)
    102.
19 J. Jeanjean, D. Delafosse and P. Gallezot, J. Phys. Chem., 83 (1979) 2761.
20 W. Romanovski, React. Kin. Catal. Lett., 4 (1976) 129.
21 Z. Maskos, J.H.C. van Hooff, J. Catal., 66 (1980) 73.
22 D.J. Elliott and J.H. Lunsford, J. Catal., 57 (1979) 11.
23 H.H. Nijs, P.A. Jacobs, J.J. Verdonck and J.B. Uytterhoeven, "Growth and
    Properties of Metal Clusters", J. Bourdon, ed., Elsevier, 1980, p. 479.
24 Ch. Minchev, V. Kanazirev, L. Kosova, V. Penchev, W. Gunsser and F. Schmidt,
    Proceedings 5th Int. Conf. Zeolites, L.V.C. Rees, ed., Heyden, London,
    Philadelphia, Rheine, 1980, p. 335.
25 J.A. Rabo, C.L. Angell, P.H. Kasai and V. Schomaker, Disc. Faraday Soc.,
    (1966) 329.
26 F. Schmidt, W. Gunsser and J. Adolph, A.C.S. Symp. Ser., 40, J. Katzer, ed.,
    1977, p. 291.
27 F. Steinbach and H. Minchev, Z. Phys. Chem. (NF), 99 (1976) 223.
28 P.A. Jacobs, H. Nijs, J. Verdonck, E.G. Derouane, J.P. Gilson and A.J.
    Simoens, J.C.S. Faraday I, 75 (1979) 1196.
29 A. Wiedenmann, F. Schmidt and W. Gunsser, Ber. Buns. Ges. Phys. Chem., 81
    (1977) 525.
30 P. Gallezot, G. Coudurier, M. Primet and B. Imelik, A.C.S. Symp. Ser., 40
    (1977) 144.
31 D. Ballivet-Tkatchenko and I. Tkatchenko, J. Molec. Catal., 13 (1981) 1.
32 D. Ballivet-Tkatchenko, G. Coudurier, H. Mozzanega and I. Tkatchenko,
    "Fundamental Research in Homogeneous Catalysis", 3, M. Tsutsui, ed.,
    Plenum, 1979, p. 257.
33 E.G. Derouane, J.B. Nagy and J.C. Vedrine, J. Catal., 46 (1977) 434.
34 T. Bein, P.A. Jacobs and F. Schmidt, this volume.
35 J.J. Verdonck, P.A. Jacobs, M. Genet and G. Poncelet, J.C.S. Faraday I,
    76 (1980) 903.

36   H. Beyer, P.A. Jacobs and J.B. Uytterhoeven, J.C.S. Faraday I, 72 (1976) 674.
37   H.K. Beyer and P.A. Jacobs, this volume.
38   L.R. Gellens, W.J. Mortier and J.B. Uytterhoeven, Zeolites, 1 (1981) 85.
39   N. Jaeger, U. Melville, R. Nowak, H. Schrübbers and G. Schulz-Ekloff, "Catalysis by Zeolites", B. Imelik et al., eds., Elsevier, 1980, p. 335.
40   R. Briend-Fauré, J. Jeanjean, D. Delafosse and P. Gallezot, J. Phys. Chem., 84 (1980) 875.
41   T.A. Egerton and J.C. Vickerman, J.C.S. Faraday I, 69 (1973) 39.
42   W. Romanovski, Rocz. Chem., 45 (1971) 427 and K.H. Bager, F. Vogt and H. Bremer, A.C.S. Symp. Ser., 40 (1976) 528 and Z. Chem., 14 (1974) 200.
43   A. Tungler, J. Petro, T. Mathé, G. Besenyei and Z. Csuros, Act. Chim. Acad. Scient. Hung., 82 (1974) 183.
44   W.J. Mortier, J. Catal., 55 (1978) 138.
45   P.A. Jacobs, W.J. Mortier and J.B. Uytterhoeven, J. Inorg. Nucl. Chem., 40 (1978) 1919.
46   P.A. Jacobs, Catal. Rev. Sc. & Eng., 1982, to be published.
47   P.A. Jacobs, M. Tielen, J.P. Linart and J.B. Uytterhoeven, J.C.S. Faraday I, 72 (1976) 2793.
48   F.O. Bravo, J. Dwyer and D. Zamboulis, Proceedings 5th Int. Conf. Zeolites, L.V.C. Rees, ed., Heyden, 1980, p. 749.
49   A. Baranski, A. Bielanski and A. Pattek, J. Catalysis, 26 s(1972) 286.
50   H.K. Beyer and P.A. Jacobs, A.C.S. Symp. Ser., 40 (1977) p. 993.
51   S.J. Gentry, N.W. Hurst and A. Jones, J.C.S. Faraday I, 75 (1979) 1988.
52   T. Kubo, H. Arai, H. Tominaga and T. Kunuzi, Bull. Soc. Chem. Soc. Japan, 45 (1972) 607, 613.
53   P. Gallezot, A. Alarcon-Diaz, J.A. Dalmon, A. Renouprez and B. Imelik, J. Catalysis, 39 (1975) 334.
54   W.J. Reagan, A.W. Chester and G.T. Kerr, J. Catalysis, 69 (1981) 89.
55   G. Bergeret, P. Gallezot and B. Imelik, J. Phys. Chem., 85 (1981) 411.
56   D. Exner, N. Jaeger and G. Schulz-Ekloff, Chem. Ing. Techn., 52 (1980) 734.
57   G. Braun, F. Fetting, C.P. Haelsig and E. Gallei, Acta Chim. Phys., 24 (1978) 93.
58   D. Hermerschmidt and R. Haul, Ber. Bunsenges. Phys. Chem., 84 (1980) 902.
59   C. Mintschev and F. Steinbach, "Molecular Sieves", J.B. Uytterhoeven, ed., Leuven University Press, 1973, p. 401.
60   M. Che, J.F. Dutel, P. Gallezot and M. Primet, J. Phys. Chem., 80 (1976) 2371.

REFLECTANCE SPECTROSCOPIC STUDY OF Ag^+, $Ag°$ AND Ag CLUSTERS IN ZEOLITES OF THE FAUJASITE-TYPE

L.R. GELLENS and R.A. SCHOONHEYDT

Centrum voor Oppervlaktescheikunde en Colloïdale Scheikunde, Katholieke Universiteit Leuven, De Croylaan 42, B-3030 Leuven (Heverlee), Belgium

ABSTRACT

Ag faujasite-type zeolites were studied after various treatments by reflectance spectroscopy. The air-dry zeolites, containing Ag^+ and small Ag particles are autoreduced upon dehydration with formation of $Ag°$, Ag_n clusters and metal particles. The degree of reduction increases with increasing Si/Al ratio. Oxidation converts the reduced particles to Ag^+ and Ag_3 clusters are formed, the concentration of which is inversely proportional to the Si/Al ratio. The hydration, dehydration and oxidation cycles are fully reversible. H_2 reduction produces $Ag°$, Ag clusters and Ag particles and subsequent oxidation restores the Ag_3 clusters and Ag^+.

INTRODUCTION

Ag zeolites are very sensible to various chemical treatments. Upon dehydration of Ag zeolites of type A and of the faujasite-type autoreduction of the Ag zeolites occurs with formation of small metal clusters inside the zeolite A framework [1-3] and Ag microcrystallites outside the zeolite [5,6]. Dehydration and subsequent oxidation is necessary to form Ag_3^{x+} clusters in faujasite-type zeolites [4]. The low temperature reduction with hydrogen is controlled by the formation of charged Ag clusters in the sodalite cages [7-9]. At high temperature the reduction is quantitative and microcrystallites are formed outside the zeolites. Oxidation converts all these Ag particles to Ag^+ in ion exchange positions.

In this work silver zeolites of the faujasite-type with a Si/Al ratio of respectively 2.9, 2.5, 1.7, 1.2 and 0.9 were studied by diffuse reflectance spectroscopy in order to characterize the various silver species formed upon high temperature dehydration, oxygen, hydrogen and water treatments.

EXPERIMENTAL SECTION

Sample preparation

NaF56 samples were supplied by Strem Chemicals, NaF70 and NaF48 by the Linde Division of the Union Carbide Corporation, NaF86 by Uetikon and NaF101 by Texaco [10]. They were exchanged for approximately 19 hours in a $AgNO_3$ solution containing twice the C.E.C. After exchange the samples were washed and air dried. All these manipulations were performed in the absence of light. In these conditions exchange is complete [11]. The chemical composition is then as follows :

AgF48 : $Ag_{48.2} Si_{143.8} Al_{48.2} O_{384} \cdot xH_2O$

AgF56 : $Ag_{55.5} Si_{136.5} Al_{55.5} O_{384} \cdot xH_2O$

AgF70 : $Ag_{69.8} Si_{122.2} Al_{69.8} O_{384} \cdot xH_2O$

AgF86 : $Ag_{86} Si_{106} Al_{86} O_{384} \cdot xH_2O$

AgF101: $Ag_{101} Si_{91} Al_{101} O_{384} \cdot xH_2O$

Pretreatment and measurements

TABLE 1

Pretreatment conditions of the silver exchanged faujasite-type samples AgF48, AgF56, AgF70, AgF86 and AgF101. The color of the sample and special remarks are included. The numbers of the AgF56 and AgF86 samples correspond to those of fig. 1 and 2.

AgF48	AgF56	AgF70	AgF86	AgF101	
grey	grey (a1)	grey	grey (a1)	grey	air dry
brown (821K)	grey-brown (a2)	grey-brown	yellow black spots (a2)	yellow black spots	dehydrated (17-20 h, $\sim$700K)
green	pale yellow (a3)	green	bright yellow (a3)	bright yellow	oxidation (O_2, >17h, 683-783K)
grey (2h)		grey (4h)	white (>12h) (a4)	white (>12h)	rehydrated
khaki		grey	bright yellow	yellow	dehydrated (>17h)
green yellow spots		pale green	bright yellow	yellow	reoxidized (O_2, >17h)
	grey (a4) 17h, 341K		green (b1) 3h, 293K		reduced (H_2)
	pale green 17h, 623K (b1)		brown (b2) 17h, 293K		reoxidized (O_2)
	pale yellow >24h, 773K (b3)		bright yellow 20h, 729K (b3)		further oxidation

Approximately 2 g of the silver exchanged zeolite were transferred in a reflectance cell [12] and subsequently pretreated. The pretreatment conditions are schematized in table 1 : before recording the spectra the samples are briefly evacuated (8.10^{-6} torr) and the amount of hydrogen in the reduction step was limited in order to obtain a partial reduction of silver. The pretreatment of the AgF56 samples is very similar to the conditions in the structural study of reference 9.

The diffuse reflectance spectra were recorded on a Cary 17 spectrophotometer equipped with a type I reflectance unit. The reference in a matching reflectance cell was $BaSO_4$. The data were processed according to the Schuster-Kubelka-Munk theory as has been previously described [2].

RESULTS AND DISCUSSION

Due to space limitations the spectra of AgF48, AgF70 and AgF101 are not reproduced, but they follow the same trends as AgF56 (Fig. 1) and AgF86 (Fig. 2). Differences will be reported in the text.

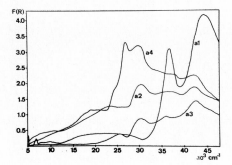

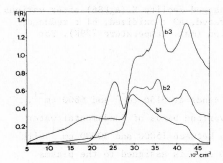

Fig. 1. Absorption spectra of silver exchanged zeolite Y (AgF56) after various pretreatments : a1 : air dry, a2 : dehydrated, a3 : oxidized, a4 : reduced, b1 : oxidized (623K), b2 : second oxidation (max. temperature 718K), b3 : third oxidation (max. temperature 773K). For details of the pretreatment, see table 1.

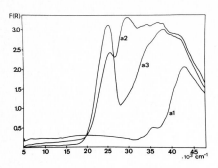

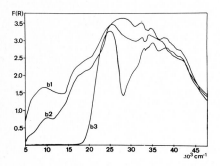

Fig. 2. Absorption spectra of silver exchanged zeolite X (AgF86) after various pretreatments : al : air dry, a2 : dehydrated, a3 : oxidized, bl : reduced, b2 : oxidized (293K), b3 : second oxidation (max. temperature 729K). For details of the pretreatment, see table 1.

a. Hydrated samples

The hydrated samples have weak narrow bands at 5150 cm^{-1} and 6800 cm^{-1}, which are respectively combination and overtone bands of fundamental water vibrations. A very broad, weak absorption between 15000 and 30000 cm^{-1} with maximum intensity between 21000 and 23000 cm^{-1} is assigned to the plasma resonance absorption of very small metal particles [13,14]. The halfwidth of the resonance peak corresponds to particles smaller than 2 nm, when compared with calculated curves [13,15-16]. This is under the detection limit of X-ray line broadening. Several bands are present in the UV region at respectively 42600–44000 cm^{-1}, a band between 35400 and 37000 cm^{-1} (only a shoulder for AgF70) and a shoulder between 33000 and 34000 cm^{-1}. The exact positions of the band maxima depend on the Si/Al ratio in that the frequencies of the main bands decrease by 1000-2000 cm^{-1} with decreasing Si/Al ratio (see fig. 1 and 2). We assign all these bands to the $^1S_o(4d^{10}) \rightarrow ^1D_2(4d^95s)$ transition of Ag^+, in agreement with the data on AgA [2] and Ag_8Y [17]. The frequencies are systematically lower than those of the free ion (46046 cm^{-1}) [18]. The high energy band is assigned to the hydrated silver ion, due to its appearance and disappearance with the presence and absence of water. The other bands involve Ag^+ on one or more sites of the zeolitic framework.

b. Dehydrated samples

Upon dehydration the 44000 cm^{-1} band disappears, a new band appears at 41000-42000 cm^{-1} and the intensity of the other bands, attributed to Ag^+ increases. They are ill-resolved because of too high Ag^+-concentrations. An intense band at 30100 ± 600 cm^{-1} is formed. The frequency decreases slightly with increasing Si/Al ratio. It is assigned to the $4d^{10}5s \rightarrow 4d^{10}5p$ transition of atomic silver located in one or more sites of the zeolite. The frequency agrees with that of gaseous Ag° and of Ag° in inert gas mactrices [18,19], but the 2 components ($^2P_{3/2}$, $^2P_{1/2}$) are not resolved in our case. The plasma resonance absorption of metal particles has vanished completely in high aluminum containing zeolites, whereas it is intensified and sharpened in high-silica zeolites. This implies larger metal particles as shown also by X-ray diffraction [9]. The dehydrated AgF48, AgF56 and AgF70 zeolites absorb in the NIR region at ∿ 10000 cm^{-1} with shoulders at the low energy side. At the same time in all samples a band at 25000 cm^{-1} (for AgF70, AgF56 and AgF48 superposed on the plasma absorption) starts to develop together with an ill-defined band at 38500 ± 1500 cm^{-1}. The 10000 cm^{-1} band is assigned to transitions within the energy levels of small intracrystalline Ag-clusters. Similar bands may be embedded under the plasma resonance absorption (fig. 1). We think to have particles of various sizes all having molecular properties because of their easy removal upon oxidation. If this assignment is true, the proposal that a frequency of 17000 cm^{-1} is the limiting low frequency absorption of such cluster does not hold [20]. The 25000-38500 cm^{-1} bands pair will be discussed in the next section.

c. Oxidized samples

The various Ag-clusters described in the previous section are quantitatively removed by oxidation except for AgF48. The 30100 cm^{-1} band of Ag° does not completely disappear in low silica zeolites, supporting the hypothesis of Ag° in site I [17]. On the other hand, the absorptions at 25000 cm^{-1} and 37000-40000 cm^{-1}, which correspond to Ag_2 or Ag_3 clusters [2,19-24], have increased. This observation agrees with the structural study [11], where the existence of Ag_3

clusters by simultaneous occupancy of sites I and I' within bonding distance is proposed. When these oxidized samples are rehydrated the spectral features of the original hydrated samples are recovered.

d. Reduced samples

Reduction with hydrogen leads to various bands in the 6000–30000 cm^{-1} region, assigned in previous sections to $Ag°$, Ag_n clusters and metal particles. The ill-defined bands in the UV region involve Ag^+ proving that silver is not yet quantitatively reduced. These observations are confirmed by the kinetic [1], the X-ray diffraction [9] and the ESR study [25] on zeolite Y (AgF56). The reduced particles can be progressively oxidized with oxygen as can be seen from fig. 1 and 2. The modulated 6000–30000 cm^{-1} absorption of reduced silver species vanishes, the 30000 cm^{-1} band of $Ag°$ being most resistent to the oxidation. Finally the spectrum of the oxidized samples is reproduced with the UV multiplet of Ag^+ and the two band system of Ag_3, when the simultaneous occupancy of site I and I' is restored [9].

CONCLUSIONS

We have been able to assign qualitatively the spectra of Ag zeolites to various Ag species : $[Ag(H_2O)_n]^+$, Ag^+, $Ag°$, $Ag_{2,3}^{x+}$, Ag_n, metal particles. For $[Ag(H_2O)_n]^+$, Ag^+ and $Ag°$ we have not observed the frequency shifts with respect to the gasphase spectra as reported by Klier et al. for Cu^+ [26]. When samples with different Si/Al ratio are compared the following trends become evident : (i) Ag zeolites are autoreduced by dehydration with formation of metal particles, Ag_n-clusters and $Ag_{2,3}^{x+}$-clusters in low Al zeolites, but only the $Ag_{2,3}^{x+}$ appears on high Al-zeolites; (ii) O_2 treatment oxidizes Ag_n and metal particles (except in AgF48). $Ag_{2,3}^{x+}$ species increase in concertation. Their concentration increases with increasing Ag content or increasing Al content of the zeolite; (iii) independently of the Si/Al ratio, reduction with hydrogen forms Ag clusters, $Ag°$ and Ag microcrystallites inside or outside the zeolite, which can be reoxidized.

ACKNOWLEDGMENTS

L.R.G. and R.A.S. acknowledge the Belgian National Fund for Scientific Research (N.F.W.O.) for research positions as respectively Research Assistant and Senior Research Associate. This research was supported by the Belgian Government (Wetenschapsbeleid, Geconcerteerde Onderzoeksakties).

REFERENCES

1 P.A. Jacobs, J.B. Uytterhoeven and H.K. Beyer, J.C.S. Faraday Trans. I, 75 (1979) 56-64.
2 L.R. Gellens, W.J. Mortier, R.A. Schoonheydt and J.B. Uytterhoeven, J. Phys. Chem., 85 (1981) 2783-2788.
3 Y. Kim, J.W. Gilje and K. Seff, J. Amer. Chem. Soc., 99 (1977) 7055-7057.
4 L.R. Gellens, W.J. Mortier and J.B. Uytterhoeven, Zeolites, 1 (1981) 11-18.
5 K. Tsutsumi and H. Takahashi, Bull. of Chem. Soc. Japan, 45 (1972) 2332-2337.
6 Kauzaki and Jasumori, J. Phys. Chem., 82 (1978) 2351-2352.
7 H.K. Beyer, P.A. Jacobs and J.B. Uytterhoeven, J.C.S. Faraday Trans. I, 72 (1976) 674-685.
8 M. Iwamoto, T. Hashimoto, T. Hamano and S. Kagawa, Bull. Chem. Soc. Japan, 54 (1981) 1332-1337.
9 L.R. Gellens, W.J. Mortier and J.B. Uytterhoeven, Zeolites, 1 (1981) 85-90.
10 The samples are indicated by F (faujasite), followed by the number of Al atoms/U.C. F48, F56 and F70 are Y-type zeolites, F86 is an X-type sample and F101 is the zeolite HP, German Patent 2734774 (1978).
11 P. Fletcher and R.P. Townsend, J.C.S. Faraday Trans. I, 77 (1981) 497-509.
12 F. Velghe, R.A. Schoonheydt and J.B. Uytterhoeven, Clays and Clay Min., 25 (1977) 375-380.
13 A. Kawabuto and R. Kubo, J. Phys. Soc. of Japan, 21 (1966) 1765-1772.
14 T. Welker and T.P. Martin, J. Chem. Phys., 70 (1979) 5683-5691.
15 L. Genzel, T.P. Martin and U. Kreibig, Z. Physik B, 21 (1975) 339-346.
16 U. Kreibig and C.V. Fragstein, Z. Physik, 224 (1969) 307-323
17 R. Kellerman and J. Texter, J. Chem. Phys., 70 (1979) 1562-1563.
18 C.E. Moore, Atomic Energy Levels, vol. III, National Bureau of Standards, Circular 467 (1958) 48-54.
19 D.M. Gruen and J.K. Bates, Inorg. Chem., 16 (1977) 2450-2453.
20 G.A. Ozin and H. Huber, Inorg. Chem., 17 (1978) 155-163.
21 S.A. Mitchell, G.A. Kenney-Wallace and G.A. Ozin, J. Amer. Chem. Soc., 103 (1981) 6030-6035.
22 G.A. Ozin, Faraday Symposia of the Chemiscal Society, no. 14, Diatomic Metals and Metallic Clusters (1980) 1-64.
23 W. Schulze, H.U. Becker and H. Abe, Chem. Phys., 35 (1978) 177-186.
24 G.A. Ozin, H. Huber and S.A. Mitchell, Inorg. Chem., 18 (1979) 2932-2934.
25 D. Hermerschmidt and R. Haul, Ber. Bunsenges. Phys. Chem., 84 (1980) 902-907.
26 J. Texter, D.H. Strome, R.G. Herman and K. Klier, J. Phys. Chem., 81 (1977) 333-338.

CHEMICAL EVIDENCE FOR CHARGED CLUSTERS IN SILVER ZEOLITES

by Hermann K. BEYER[1] and Peter A. JACOBS[2]

[1]Central Research Institute of Chemistry of the Hungarian Academy of Sciences, P.O.Box 17, Budapest (Hungary)

[2]Centrum voor Oppervlaktescheikunde en Colloidale Scheikunde, Katholieke Universiteit Leuven, De Croylaan 42, B-3030 Leuven (Heverlee) (Belgium)

ABSTRACT

Hydrogen uptake by silver chabasite was followed volumetrically in the isothermal and temperature programmed mode. This reduction occurs in three distinct steps. Only after the final step, silver crystals were found by X-ray line broadening at the external surface of the zeolite crystals. After the first two steps a distinct but constant degree of reduction was obtained, independently of the initial Ag^+ loading. Combined with EPR measurements, this allows to conclude that when the reduction progresses two clusters of the following composition are formed: Ag_4^{2+} and Ag_3^+.

The reduction in AgY and Ag-mordenite is similar and also leads to intermediate formation of charged clusters, which are thermally very stable and cannot be reoxidized easily with oxygen.

INTRODUCTION

Zeolite supported metal particles are in general prepared by ion-exchange of the corresponding transition metal ion into the zeolite structure, followed by reduction of the ions with molecular hydrogen. Under mild reduction conditions several authors claim to have obtained charged clusters using silver zeolites.

From the kinetics of hydrogen consumption of AgY zeolites (ref. 1) and mordenites (ref. 2), Beyer et al. concluded that uncharged silver clusters formed at low reduction temperatures, associate with silver ions and form charged silver clusters with an average composition of Ag_3^+ and Ag_5^+, respectively. In the mean time, Gellens et al. (ref. 3) by X-ray methods, were able to detect charged silver clusters in the sodalite cages of partly reduced AgY zeolites. Moreover, these authors showed that such clusters resist remarkably thermal sintering in an inert atmosphere.

In partly reduced and self-reduced A-type zeolites, Kim et al. (ref. 4,5) also proposed the formation of charged silver clusters on the basis of single crystal X-ray structure determinations. Hermerschmidt and Haul (ref. 6) using EPR spectroscopy, located these clusters in the sodalite units of Ag-zeolites and were able to detect at room temperature the decay of these species and the subsequent formation of uncharged particles.

In the present work chemical evidence is collected which supports the existence of charged clusters in Y, mordenite and chabasite-type zeolites. Emphasis
will also be on the mechanism of formation and chemical stability of these
clusters.

EXPERIMENTAL

Materials. The AgY zeolites and Ag-mordenites (MOR) were prepared by ion exchange with 0.1 M $AgNO_3$ solutions of NaY from Union Carbide Corp. and Zeolon-
-100 from Norton Comp., respectively. The starting material for Ag-chabasites
(CHA) was natural chabasite from Dunabogdány (Hungary). The unit cell composition of the different materials prepared by this procedure is given in Table 1.

TABLE 1. Unit cell composition of the Ag-zeolites

sample notation	unit cell composition
$Ag_{0.90}$ Na-CHA	$Ag_{3.31}$ $Na_{0.36}$ $[Al_{3.67}$ $Si_{8.33}$ $O_{24}]$
$Ag_{0.75}$ Na-CHA	$Ag_{2.75}$ $Na_{0.92}$ $[Al_{3.67}$ $Si_{8.33}$ $O_{24}]$
$Ag_{0.68}$ Na-CHA	$Ag_{2.50}$ $Na_{1.17}$ $[Al_{3.67}$ $Si_{8.33}$ $O_{24}]$
$Ag_{0.51}$ Na-CHA	$Ag_{1.87}$ $Na_{1.80}$ $[Al_{3.67}$ $Si_{8.33}$ $O_{24}]$
$Ag_{0.70}$ Na-Y	$Ag_{38.5}$ $Na_{16.2}$ $[Al_{57.7}$ $Si_{137.3}$ $O_{384}]$
$Ag_{0.87}$ Na-Y	$Ag_{47.6}$ $Na_{7.1}$ $[Al_{57.7}$ $Si_{137.3}$ $O_{384}]$
$Ag_{0.55}$ Na-MOR	$Ag_{4.22}$ $Na_{3.90}$ $[Al_{7.62}$ $Si_{40.38}$ $O_{96}]$

Before the reduction experiments, the Ag-zeolites were dehydrated by temperature programmed heating in vacuo up to 673 K (4 K/min heating rate).

Methods. Temperature programmed reduction (TPR) and reoxidation (TPO) as
well as isothermal hydrogen uptake measurements, were performed in a recircu-
lating system by measuring pressure changes (ref. 1). X-ray diffractograms of
the reduced and reoxidized samples were taken on a Philips PW 1130/00 diffractometer. The size of silver particles was calculated from the broadening of
the [111] diffraction line (ref. 7).

RESULTS AND DISCUSSION

1. Step-wise reduction of AgNa-CHA by molecular hydrogen

Isothermal uptake curves

The uptake of molecular hydrogen in isothermal conditions but at different
temperatures for $Ag_{0.51}$Na-CHA is shown in Fig. 1. The set of uptake curves
determined this way shows that the reduction progresses in clearly distingui-

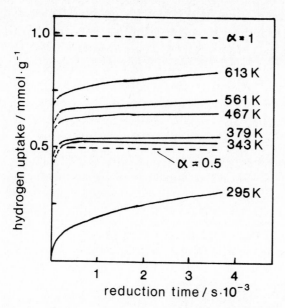

Figure 1. Uptake of molecular hydrogen by $Ag_{0.51}Na$-CHA at different temperatures (The horizontal dashed line corresponds to the theoretical uptake).

shable steps. The first step is already finished around 343 K, since further increase of the reduction temperature up to 371 K hardly enhances the amount of hydrogen taken up. Under these reaction conditions the process is very fast. Therefore, kinetic curves for this reduction step can only be measured at temperatures below 420 K. A second reduction step is completed already around 467 K (degree of reduction equals about 0.65). The figure also shows that complete reduction of the remaining Ag^+ can only be performed in severe conditions.

A similar multi-step reduction behaviour was observed earlier for Ag-Y and Ag-MOR zeolites (ref. 1,2). For these zeolites this step-wise reduction was explained in terms of site preference of the cations and of the formation of charged clusters. Under mild reduction conditions, it was suggested that first charged clusters are formed (low temperature reduction mechanism). During the second step these clusters are further reduced, while it requires still higher reduction temperatures to attack Ag^+ ions in the hidden sites (sites I in case of Y and sites II in the side-pockets of the MOR structure). Whether a similar explanation holds in the present case will be discussed in the next paragraphs.

TPR curves

The TPR curves of Ag-CHA with different ion exchange degree, shown in Fig. 2, reflect the same phenomena.

In each case, three reduction steps can be well distinguished. The first step proceeding already at room temperature is completed in the TPR conditions at about 370 K. Independently of the ion-exchange degree always about 50% of the Ag^+-ions present are involved in this reduction step. Concomitant colour changes from white to ochreous (yellowish-brown) occur during this step. Depending on the silver lattice ion concentration, the second reduction step proceeds in the temperature range from 370 to 470 K. During this step the sample turns its colour to olive-green. After this step, roughly 2/3 of the Ag^+ ions originally present are reduced. Complete reduction of the Ag-CHA samples takes only place in the temperature range of 620 to 750 K.

Accurate values of the reduction degrees reached after the first (α_1) and second (α_2) step are given in Table 2 for Ag-CHA samples with various degrees of Ag^+ exchange. The very characteristic colours as well as the constant degree of reduction obtained after each step, independently of the original Ag^+ loading of the CHA-sample, point to the existence of two distinct types of charged clusters.

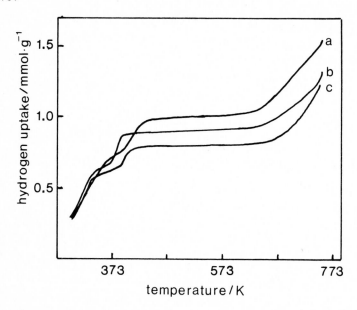

Figure 2. Temperature programmed reduction of (a), $Ag_{0.90}$Na-CHA, (b), $Ag_{0.75}$Na-CHA and (c), $Ag_{0.68}$Na-CHA by molecular hydrogen (heating rate 4K/min).

EPR signals which eventually can be ascribed to unpaired electrons in such clusters, have not been detected, neither during the first nor during the second step. This may suggest that the index m does not represent an odd number. Therefore m most probably is equal to 2 or 4, which indicates that this charged cluster must be located in the big cavities of the chabasite structure.

TABLE 2. Reduction degree of AgNa-CHA after the first (α_1) and second step (α_2) in the hydrogen-uptake curves.

Ag^+/UC	α_1	α_2
3.30	0.50	0.64
2.75	0.50	0.66
2.49	0.49	0.65
1.85	0.53	0.65
sample colour	ochreous	olive-green

The absolute values for α_1 and α_2 suggest that clusters of the following composition should be formed:

$(Ag_2^+)_m$: during the first step

$(Ag_3^+)_n$: during the second step

Clusters with an average composition of Ag_3^+ (if n=1) can be easily accommodated in the chabasite structure, Ag^+ being located at SI and Ag^O at SII. It is obvious, that the 2 Ag^O in the Ag_3^+ cluster are on sites SII, belonging to 2 neighbouring CHA-cages, sharing a DGR unit, (an hexagonal prism).
A similar occupation of sites I and I∷ by clusters consisting of three Ag ions was also advanced for the faujasite structure (ref. 8). This would imply that upon increasing reduction of Ag-CHA, the initially formed $(Ag_2^+)_m$ clusters are redissolved and relocate themselves as Ag_3^+, into coordinatively more favourable positions. Another implication is that only a single Ag_3^+ cluster can be accommodated per unit cell. Just as in the present case, the existence of charged Ag clusters of different composition was reflected by a distinct sample colour (ref. 9).

Mechanism of the first reduction step in Ag-CHA

As can be seen in Fig. 3, kinetic curves of hydrogen-uptake measured in the temperature range of 293 to 313 K, fit the expression:

$$d[Ag^+]/dt = k \cdot [Ag^+]^2 \cdot P_{H_2}^{1/2}$$

$[Ag^+]$ represents the actual concentration of irreduced and isolated ions but does not include the residual charge of the $(Ag_2^+)_m$ clusters. The kinetic equation is accounted for by the following mechanism provided the third equation is rate-determining:

$$H_2 \rightleftarrows 2\,H\cdot$$

$$H\cdot + Ag^+ \rightleftarrows [Ag^O H^+]$$

$$[Ag^O H^+] + Ag^+ \rightarrow Ag_2^+ + H^+$$

This mechanism requires that the dissociation of hydrogen is homolytic and that the mobility of Ag^+ ions in the structure determines the reduction rate. The value of the activation energy for the first reduction step (49 kJ mol^{-1}) is of an acceptable agreement with the value reported for AgY (40 kJ mol^{-1}, ref. 1).

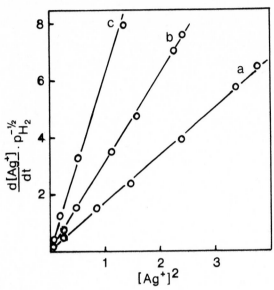

Figure 3. Linearized representation of the reduction kinetics of the initial step for $Ag_{0.90}$Na-CHA, at (a), 293 K, (b), 303 K, and (c), 313 K.

2. Sintering of charged silver clusters in zeolites

Thermal sintering

It was reported earlier (ref. 1,2) that after complete reduction of Ag-Y and Ag-MOR between 620 and 640 K, silver crystallites about 10 to 25 nm in size are formed on the outer surface of the zeolite crystals. However, these crystallites are not detectable in partially reduced zeolites (Table 3).

This unexpected behaviour of partially reduced silver zeolites can only be explained assuming the formation of charged clusters. It is obvious that positively charged silver particles are retained in the lattice by the strong electrostatic field of the negatively charged framework. In this way they are prevented from migration and further agglomeration. Moreover, a stoichiometric relation seems to exist between the number of unreduced Ag^+ ions necessary for the stabilization of the reduced part and the number of secondary building units per unit cell of the faujasite structure. It can be seen from Table 3 that silver appears on the outer surface when the unit cell contains less than 16 Ag^+ ions. On the other hand the unit cell of faujasite contains 16 hexagonal prisms and 8 cubooctaeders, respectively.

It was already reported that exactly 16 Ag^+ ions per unit cell were involved in the high temperature reduction step of Ag-Y zeolite (ref. 1). At that time the conclusion seemed justified that these ions were located in SI positions and had to migrate into cubooctaedra for reduction.

TABLE 3. Reduction conditions and characterization of the silver particles

sample	reduction				heat[2]	large Ag particles	
	temp. K	time min	degree %	$Ag^{+[1]}$ per UC	treatm. min	rel.[3] conc.%	diam. nm
Ag$_{0.70}$Na-Y	363	90	45.8	20.9	60	0	-
	445	80	63.6	14.0	60	16	17
	450	75	63.6	14.0	-	16	17
	597	60	99.2	0.3	60	139	28
	629	60	100.0	0.0	-	100	29
Ag$_{0.87}$Na-Y	363	60	51.0	24.3	60	0	-
	423	60	61.5	18.1	60	0	-
	493	60	68.8	14.8	-	32	17
	543	120	94.7	2.4	-	44	21
	543	60	85.1	7.1	60	36	22
	623	60	100.0	0.0	-	100	33
Ag$_{0.55}$Na-MOR	435	80	64.5	1.5	60	0	-
	437	80	66.1	1.4	-	0	-
	536	60	75.8	1.0	60	15	18
	533	60	75.8	1.0	-	8	12
	640	60	100.0	0.0	60	138	25
	644	60	100.0	0.0	-	100	25

[1] unreduced silver ions per unit cell after reduction

[2] heat treatment in vacuo after reduction at 620 K (Ag-Y) and 640 K (Ag-mordenite), respectively

[3] from the peak area of the (111) reflection

In case of zeolite Y, the table shows that about 16 unreduced Ag^+ ions per unit cell have to remain in the structure to prevent sintering of the metal to the outer surface. This is not a mere result of the low reduction temperature used, since even after heat treatment in vacuum between 620 and 640 K, external silver is not formed. However, the hydrogen-uptake curves of ref. 1 (Fig. 2), when they are interpreted as for Ag-CHA, also before the final reduction step show a constant degree of reduction, irrespective of the initial Ag^+ content. This value amounts approximately to 0.60, corresponding to a Ag_5^{2+} cluster, which necessarily has to be located in the sodalite cages. Since only 8 sodalite cages are present per unit cell, it is clear that for such a cluster a maximum number of 16 charges will be very difficult to reduce.

Reoxidation behaviour of charged silver clusters

It was found that after total reduction, silver zeolites can be completely reoxidized by oxygen at temperatures above 550 K (ref. 1,2). Partially reduced Ag-zeolites, however, show a different reoxidation behaviour. Even at temperatures above 550 K only part of the reduced silver can be reoxidized. As can be seen from Table 4, this portion increases with decreasing reduction degree. This unusual and unexpected reoxidation behaviour is also in favour of the charged cluster hypothesis.

TABLE 4. Reoxidation of $Ag_{0.87}$Na-Y reduced under different conditions

	isothermal reduction			reoxidation[a]		
temp. K	time min	degree %	H_2 consumption mmol g^{-1}	O_2 consumption mmol g^{-1}	degree %
362	60	56.7	0.804	0.196	48.8
433	60	64.9	0.921	0.324	70.4
539	100	89.9	1.276	0.570	89.3
623	60	100.0	1.429	0.712	99.7

[a]TPO up to 610 K.

In the same way, it was found that the charged silver clusters in the mordenite pores and the chabasite cages are also very resistent towards reoxidation. Therefore it can be concluded that this phenomenon is not of steric but probably of an electrostatic and coordinative origin.

ACKNOWLEDGMENT

The authors acknowledge a grant in the frame of a joint project between the Hungarian International Cultural Institute and the Belgian National Fund of Scientific Research. P.A.J. is grateful to the Belgian National Fund of Scientific Research for a permanent research position as "Onderzoeksleider". The skillful technical assistance of Mrs. I.Szaniszlö (Budapest) is appreciated.

REFERENCES

1 H.Beyer, P.A.Jacobs and J.B.Uytterhoeven, JCS Faraday Trans I, 72 (1976) 674-685.
2 H.K.Beyer and P.A.Jacobs, ACS Symp. Ser., 40 (1977) 493-503.
3 L.R.Gellens, W.J.Mortier and J.B.Uytterhoeven, Zeolites, 1 (1981) 85-90.
4 Y.Kim, J.W.Gilje and K.Seff, J. Am. Chem. Soc., 99 (1977) 7055-7057.
5 Y.Kim and K.Seff, J. Am. Chem. Soc., 100 (1978) 175-180.
6 D.Hermerschmidt and R.Haul, Ber. Bunsenges. Phys. Chem. 84 (1980) 902-907.
7 W.B.Innes, in R.B.Anderson (Ed.), Experimental Methods in Catalytic Research, Academic Press, New York, 1968, p. 84.
8 L.R.Gellens, W.J.Mortier and J.B.Uytterhoeven, Zeolites, 1 (1981) 85-90.
9 L.R.Gellens, W.J.Mortier, R.A.Schoonheydt and J.B.Uytterhoeven, J. Phys. Chem., 85 (1981) 2783-2788.

UV/VIS TRANSMISSION SPECTROSCOPY OF SILVER ZEOLITES. I. DEHYDRATION AND
REHYDRATION OF AgA AND AgX

H.G. KARGE

Fritz-Haber-Institut der Max-Planck-Gesellschaft, Faradayweg 4-6,
1000 Berlin 33 (Federal Republic of Germany)

ABSTRACT

Dehydration and rehydration of highly exchanged AgA and AgX zeolites have
been studied by means of UV/VIS transmission spectroscopy. A yellow form of
AgX exhibits a band around 440 nm, disappearing on interaction with O_2. In the
case of AgA three distinct stages of hydration can be distinguished, characterised by absorbances at 455 nm, 535 nm, and 585 nm. They can be reversibly
converted into each other. These spectral changes are discussed in connection
with the so-called auto-reduction of transition metal zeolites and the formation
of metal clusters in such systems.

INTRODUCTION

In an early study by Ralek et al. [1] it was reported that the silver form
of zeolite A exhibits a red colour after dehydration at 623 K. Upon contact
with water vapour the AgA sample discoloured. Similar observations were reported by Seff [2]. The behaviour of AgA was investigated in more detail by
Jacobs et al. [3] as well as by Gellens et al. [4], using several physical and
physicochemical methods, in particular temperature-programmed desorption of O_2,
adsorption of CO, H_2, and H_2O on AgA, X-ray structure analysis and UV/VIS reflectance spectroscopy. On the basis of their results, these authors developed
a two-step mechanism for auto-reduction. According to that mechanism, up to 8%
of Ag^+ should be reduced during the heating of the AgA sample to 673 K in
flowing helium or vacuum, thereby oxidising an equivalent amount of lattice O^{2-}
-ions and releasing molecular oxygen. The above authors correlated the yellow
and red forms of AgA with the first and second steps in auto-reduction, which
should result in the formation of one and two $Ag^+-Ag°-Ag^+$ per sodalite cage,
respectively [3,4]. Moreover, in ref. [3] the adsorption of two molecules H_2O
per Ag_3^{2+} species was claimed. The correlation between colour changes and interaction with H_2O remained, however, unclear. In the present paper the dehydration and rehydration processes were followed by means of UV/VIS transmission
spectroscopy in an attempt to clarify whether or not the different colours
correspond to different stages of auto-reduction. For comparison, we have included AgX in the present study and also looked for similar colour changes and
adsorption effects.

EXPERIMENTAL

The AgA and AgX samples were identical with those used in the work of Hermerschmidt and Haul [5]. They were analysed with the help of atomic absorption spectroscopy. The composition of the water-free zeolites corresponded to the formulas

$$Ag_{1.2}Na_{0.7}Al_{11.9}Si_{12}O_{48} \text{ (AgA)} \text{ and } Ag_{73.7}Na_{12.3}Al_{86}Si_{106}O_{384} \text{ (AgX)}.$$

After ion exchange and dehydration/rehydration the crystallinity of the samples was checked by X-ray analysis. No loss of crystallinity was observed. Spectroscopic investigations were carried out on self-supporting wafers (thickness: 5-10 mg/cm^2) which were pressed from the zeolite powder under 2×10^8 Pa. Prior to use the wafers were slowly heated and calcined in flowing oxygen (AgA: 144 h at 773 K, AgX: 100 h at 673 K, 20 h at 773 K). The spectra were run with an UV/VIS spectrometer 556 (Perkin-Elmer), using the double beam transmission mode. The cell could be joined to a high vacuum and gas dosing system through an all-metal UHV valve and a stainless steel bellows. Thus, the sample could be pretreated at higher temperatures outside the spectrometer cell compartment. For recording the spectra, the cell could be lowered into the cell compartment without disconnecting it from the high vacuum system. Since the cell was equipped with NaCl windows, IR measurements (Perkin-Elmer 325 spectrometer) were possible with the very same wafers used for UV/VIS experiments. The amount of adsorbed water was determined via the pressure drop measured by a capacitance manometer (BARATRON MKS) when water vapor was expanded from a known volume into the UV/VIS cell.

RESULTS

The fresh AgA samples remained white after degassing at room temperature. During the pretreatment in flowing oxygen the colour turned to brick red. Cooling in contact with air changed the colour again, viz. to light yellow. Static oxidation, i.e. 10 - 20 h calcination under 2.7×10^4 Pa oxygen, also gave the brick red colour. AgX samples were yellow at the end of the pretreatment in flowing oxygen, but white after cooling in air.

The AgA, that was not pretreated, exhibited a spectrum without any structure in the visible region, as can be seen from Fig. 1 ("white", spectrum a). This spectrum was not reproduced after oxygen treatment and subsequent contact with the moisture of ambient air. On the contrary, a rather sharp absorption edge occurred at 455 nm ("light-yellow", spectrum b). Heating at 673 K (under 3.5×10^4 Pa oxygen) shifted this edge to longer wavelengths ("brick red", spectrum c). Subsequent interaction with water vapour (ca .10 Pa and 5.0×10^2 Pa) at room temperature reversed this shift: see spectrum c (red) through d (orange) to e (yellow). A second calcination at 673 K led again to

the brick red state (f) and the cycle could be performed repeatedly. No such
sharp absorption edge was observed with identically treated NaA, the starting
material for the AgA sample.

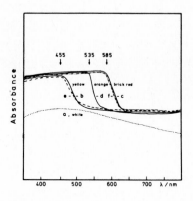

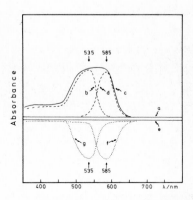

Fig. 1. Transmission spectra of AgA   Fig. 2. Difference spectra of AgA
        (for details see text)   (for details see text)

The effect of dehydration and rehydration became even more obvious when difference spectra were taken, i.e. by subtracting spectrum b of Fig. 1 from subsequent spectra. Thus, the following spectra of Fig. 2 demonstrate the spectral
differences which appear when the fully hydrated (light yellow) AgA sample is
dehydrated and subsequently rehydrated: a) the base line spectrum a was obtained
after electronic subtraction of spectrum b of Fig. 1 from the spectrum of a
pretreated sample; b) after subsequent degassing at room temperature (1 h,
1.3×10^{-4} Pa) the band of spectrum b appeared, having a maximum at 535 nm
(sample was orange); c) upon heating to 673 K under 2.7×10^{4} Pa O_2 the broad
band of spectrum c developed (sample was red); d) the band c, in fact, was com-
posed of two overlapping bands, i.e. the band d could be scanned separately,
provided that spectrum b was also subtracted after the 673 K treatment. On
contact with 10 Pa H_2O vapor, the band d disappeared. At higher H_2O pressures
(500 Pa) band b was also eliminated; e) the base line e was obtained by subtraction of the complete spectrum c after the 673 K treatment; f) correspon-
dingly, if after scan of spectrum c the sample was first contacted with 10 Pa
H_2O vapor and spectrum c then subtracted, the "negative" band f appeared; g)
admittance of higher H_2O pressure (500 Pa), followed by subtraction of spectrum
f, resulted in spectrum g. The adsorption measurements showed that about
2.2 mol H_2O/kg AgA (or 5.8 molecules H_2O per unit cell) were taken up before

the band around 535 nm was completely eliminated. This corresponds to about 50% of the Ag present in the AgA sample (4.2 mol Ag/kg AgA or 11.2 Ag per unit cell). The decrease in absorbance at 535 nm, however, was not directly proportional to the amounts of H_2O adsorbed.

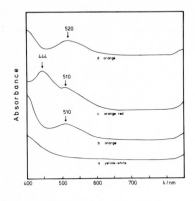

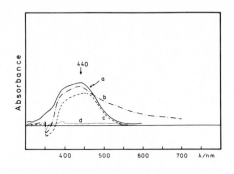

Fig. 3. Effect of oxygen on AgA
(for details see text)

Fig. 4. Difference spectra of AgX
(for details see text)

When a fully hydrated AgA sample (Fig. 3, spectrum a) was degassed at 273 K and 1.3×10^{-4} Pa (Fig. 3, spectrum b; colour orange) and subsequently heated at 723 K under 2.7×10^3 Pa O_2, the spectra (recorded in the presence of O_2) showed a distinct band at 444 nm as well as a shoulder around 510 nm (Fig. 3, spectrum c). Upon O_2 desorption at 300 K the 444 nm band was completely eliminated and the second band shifted to higher wavelengths (Fig. 3, spectrum d). This process was also entirely reversible.

The complete or partial hydration of the yellow or orange form of AgA was indicated by a strong $\delta(H_2O)$ band at 1650 cm^{-1} in the IR region. Subsequent CO admission did not lead to any CO adsorption at Ag^+; very weak CO – Ag^+ bands, or none at all, appeared around 2170 cm^{-1} after dosing of 6.7×10^3 Pa CO. The VIS spectra also showed no change due to the interaction between CO and AgA. In contrast, CO adsorption on completely dehydrated AgA (brick red form, no $\delta(H_2O)$ band in the IR) resulted in the formation of a very intense band at 2170 cm^{-1}. The maximum absorbance of this IR band was solely defined by the brick red state (characterised by the VIS double band at 535 and 585 nm), i.e. it was invariant against different pretreatments (oxidation, H_2O loading) prior to the final complete dehydration (473 K, 10^{-4} Pa). During CO adsorption onto brick red AgA, however, a concomitant colour change to brownish-yellow and a general increase of the absorbance in the VIS region above 600 nm were observed.

When an AgX sample, which had been calcined at 673 K and 2.7×10^4 Pa O_2 and cooled down in contact with air, was subsequently heated in high vacuum to 373 K (5 K/min.) the colour turned from white to yellow. The VIS spectrum of the yellow sample exhibited a broad band, showing a maximum at 440 nm and a shoulder at 390 – 400 nm (Fig. 4, band a). Here, admittance of H_2O vapour (6.7×10^2 Pa at room temperature) did not appreciably change the spectrum (Fig. 4, band b). Similarly, adsorption of CO_2 could not eliminate the band. Contact with O_2 (2.7×10^4 Pa) at room temperature, however, weakened the band and, finally, O_2 interaction at 423 K removed it completely; the sample became white again (Fig. 4, band c and d, respectively).

Adsorption of CO (6.7×10^3 Pa) did not affect the visible spectrum of the white and yellow form of AgX. However, the IR spectrum of the very same sample showed an intense line at 2174 cm^{-1}. This is ascribable to CO adsorbed at Ag^+ ions [6]. The intensity of the 2174 cm^{-1} band was exactly the same, independent of whether the AgX was in the white or yellow form.

DISCUSSION

In the present study three different forms of AgA were observed, depending on the degree of dehydration (see Fig. 1 and 2). These states are characterised by distinct bands in the visible region with maxima at 455 nm (yellow form), 535 nm (orange form), and 585 nm (brick red form). The latter appear at longer wavelengths than the main bands observed in Ref. [4] by reflectance spectroscopy (446.4 nm and 512.8 nm in the case of the yellow and red form, respectively). In our experiments, a band at 444 nm appeared only in the presence of O_2 (Fig. 3, spectrum c). Therefore, this band is tentatively ascribed to chemisorbed oxygen. Admittedly, the band of the yellow form (455 nm) is less precisely determined in our experiments, since no difference spectra could be recorded. The form of AgA, however, which is reached at higher temperatures gives rise to two overlapping bands (at 535 and 585 nm) having similar intensities. Without sufficient resolution, the observed position of the resulting maximum will depend on the intensity ratio of the two bands.

The transitions between the three different forms of AgA are completely reversible, depending only on the interaction with water vapour at room temperature (Figs. 1 and 2). Therefore it seems unlikely that the change from the yellow form to the red form is caused by an increasing extent of Ag^+ reduction. Rather, the absorption bands (455 nm, 535 nm, 585 nm) must be ascribed to three different hydration states of the AgA zeolite.

On the basis of the present observations it seems likely, however, that the hydration in question occurs on complexes containing reduced Ag. As can be seen in Fig. 1, in no case does the interaction with H_2O restore the original, completely untreated AgA sample (Fig. 1, spectrum a). It is thus conceivable that the strong background absorption remaining below 450 nm (Fig. 1, spectrum b) is due to reduced $Ag°$, which could have formed during the high temperature pretreatment, in agreement with the auto-reduction mechanism [3]. Such auto-reduction may even proceed in the presence of O_2, due to irreversible changes of the anion lattice (formation of oxygen defects and true Lewis sites, respectively [7]). The CO adsorption experiments suggest that there exists a strong interaction between Ag^+ and H_2O in AgA. Only after complete removal of the solvate water does bonding of CO onto Ag^+ become possible. Similarly, formation of the $Ag^+ - Ag° - Ag^+$ complexes seems to require sufficient dehydration. Only removal of the water molecules would permit adjacent Ag^+ ions to approach $Ag°$ atoms, forming Ag_x^{n+} clusters ($x > n$). Hydration, in turn, would dissolve the complex. The amount of adsorbed H_2O necessary to eliminate the band at 535 nm was significantly higher than the fraction of Ag^+ which should be involved in the formation of $Ag^+ - Ag° - Ag^+$ clusters (ca. 16% of the total Ag, see Ref. 3). Obviously, a (somewhat varying) fraction of the admitted H_2O was adsorbed on Ag^+ which were not adjacent to reduced silver atoms. Due to the high background absorption below 350 nm, atomic $Ag°$ could not be detected in our experiments; it should give rise to a band around 306 nm [8].

Surprisingly, the band at 535 nm is not simply shifted to higher wavelengths under more severe dehydration conditions at 673 K. This band remains in fact at more or less the same intensity, and the band at 585 nm develops in addition. The reason may be that after complete removal of water at 673 K, adjacent Ag_x^{n+} - clusters can interact, giving rise to additional electronic transitions.

In contrast to the 535 nm and 585 nm bands observed for AgA, the band of yellow AgX (showing up in the 400 - 450 nm region) could not be eliminated by H_2O admittance, but disappeared during oxygen treatment at 423 K. Further investigations are required in order to clarify whether this effect is due to chemisorption of O_2 on Ag species or to the reversibility of Ag^+ auto-reduction. We can, however, conclude from the constant intensity of the 2174 cm^{-1} band that the number of Ag^+ ions accessible to CO (i.e. located in the supercages) is the same in the white and yellow form of AgX. Thus, if any auto-reduction of Ag^+ occurs during the heating of the AgX sample at 373 K in high vacuum, then it must take place in the small cavities, from which CO is excluded.

ACKNOWLEDGEMENT

The author thanks Mrs. U. Köckeritz and Mrs. E. Popović for their valuable technical assistance. Thanks are also due to Prof. Dr. J.H. Block for helpful discussions. The financial support of this work from the Deutsche Forschungs-gemeinschaft (DFG) is gratefully acknowledged. The author is indebted to Dr. D. Hermerschmidt and Prof. Dr. R. Haul for kindly supplying the silver zeolites.

REFERENCES

1　M. Ralek, P. Jirů, O. Grubner, and H. Beyer, Coll. Czechoslov. Comm., 26 (1961) 142.
2　Y. Kim and K. Seff, J. Am. Chem. Soc., 100 (1978) 6989.
3　P.A. Jacobs, J.B. Uytterhoeven, and H. Beyer, J. Chem. Soc., Faraday Trans. I, 75 (1979) 56.
4　L. Gellens, W.J. Mortier, R.A. Schoonheydt and J.B. Uytterhoeven, J. Phys. Chem., 85 (1981) 2783.
5　D. Hermerschmidt and R. Haul, Ber. Bunsenges. Phys. Chem., 84 (1980) 902.
6　H. Beyer, P.A. Jacobs and J.B. Uytterhoeven, J. Chem. Soc., Faraday Trans. I, 72 (1976) 674.
7　P.A. Jacobs and H.K. Beyer, J. Phys. Chem., 83 (1979) 1174.
8　R. Kellerman and J. Texter, J. Chem. Phys., 70 (1979) 1562.

ADSORPTION AND DECOMPOSITION OF IRON PENTACARBONYL ON Y ZEOLITES

Th. BEIN[1,3], P.A. JACOBS[2] and F. SCHMIDT[1]

[1]Institut für Physikalische Chemie der Universität Hamburg, Laufgraben 24, D-2000 Hamburg 13 (F.R.G.)

[2]Centrum voor Oppervlaktescheikunde en Colloïdale Scheikunde, Katholieke Universiteit Leuven, De Croylaan 42, B-3030 Leuven (Heverlee) (Belgium)

[3]Present address : 2

ABSTRACT

The adsorption isotherms of $Fe(CO)_5$ on NaY and HY zeolites obtained in McBain balances show micropore adsorption, the saturation at $p/p_o = 0.5$ being 39 and 42 % per dry wt, respectively. IR results indicate a restricted mobility of the encaged complex. Nevertheless it can thermally be desorbed to a great extend in vacuum.

For the first time, well distinguishable decomposition phases of zeolite-adsorbed $Fe(CO)_5$ are found by thermogravimetric analysis. These phases are associated with species bearing 2(4) and 1/4(1) CO ligands per Fe in the case of NaY(HY). New evidence is found for the intermediate $Fe_3(CO)_{12}$. The slow decomposition reaction in inert atmosphere is completed already between 70 and 90°C, providing an iron content of 10.5 ± 0.5 wt %.

INTRODUCTION

With respect to the industrial importance of iron catalysts and the still not entirely understood particle size effect in catalysis, it is desirable to dispose of model catalysts with variable, narrow particle size distribution. Zeolites have proved to be suitable supports for metals (ref. 1,2) and to behave as model catalysts. Our aim is therefore to obtain Fe(0) containing zeolites with narrow particle size distributions. Reduction of Fe(II) exchanged Y-type zeolites was found to be impossible with H_2 (ref. 3,4), whereas reduction with sodium vapor resulted in highly dispersed iron metal (ref. 5-8).

With regard to the difficult procedures to be used in these methods, the decomposition of Y-zeolite adsorbed iron pentacarbonyl was chosen as an alternative.

Thermal decomposition of this complex has already been used to prepare dispersed supported iron (ref. 9), while recently it has been applied to $Fe(CO)_5$ loaded HY zeolite (ref. 10-12). In addition, decomposition by UV light was reported to provide a highly dispersed iron phase in the HY zeolite.

In the former studies, few quantitative details are given with regard to the parameters which govern adsorption and decomposition of the complex. In order to arrive at a quantitative understanding of these processes, the adsorption and decomposition behaviour of iron pentacarbonyl in NaY and HY zeolite has been studied by means of gravimetric, thermogravimetric and IR-spectroscopic methods.

EXPERIMENTAL

Materials

Synthetic NaY with Si/Al = 2.46 was from Strem Chemicals. It was treated with 0.1 M NaCl solution ot remove possible cation deficiencies, washed and air dried, and stored over saturated NH_4Cl solution. The NH_4Y form was obtained by conventional ion exchange. Before loading with iron carbonyl, both zeolites were degassed in situ at 450°C for about 12 hrs at 10^{-5} mbar, at a heating rate of 2°C/min.

Iron pentacarbonyl from Ventron (99.5 %) was cold distilled in the dark and stored over molecular sieve 5A. The zeolite samples for the McBain and IR measurements were loaded with the carbonyl as follows. The frozen carbonyl was outgassed in vacuum and allowed to warm up until the desired pressure was reached. All procedures with $Fe(CO)_5$ were performed in the dark, whereas the weight measurements at the McBain balance have been carried out in weak red light.

Methods

Adsorption isotherms were obtained in a McBain balance with calibrated quartz spring, with a precision of ± 0.5 %. The pressure was measured with a Bell & Howell pressure transducer BHL-4100-01, which is linear within ± 0.5 % up to 750 mbar.

Infrared spectra were taken with a Perkin Elmer 580B spectrometer from 4000 to 1200 cm^{-1} (resolution 2 cm^{-1}) using a quartz cell with 80 mm path length and equipped with CaF windows of 3 mm thickness. The zeolite was pressed at 1 ton/cm^2 to selfsupporting films of ca. 5 mg/cm^2. All treatments were performed in situ in the IR cell.

Thermogravimetric measurements were done on a Mettler Thermoanalyzer 2 under He purge, mostly in the 10 mg range. Samples of 5 to 50 mg zeolite were outgassed by heating at 2°C/min up to 450° in a quartz oven and loaded at 20°C in a stream of dry helium containing ca. 4 mbar $Fe(CO)_5$. The flow rate of this stream was 2.8 1/h.

RESULTS AND DISCUSSION

1. Adsorption isotherms of $Fe(CO)_5$ on Y-zeolite

Comparison of the $Fe(CO)_5$ adsorption isotherms on NaY and HY shows rather similar behaviour (Fig. 1). The major uptake occurs at very low partial

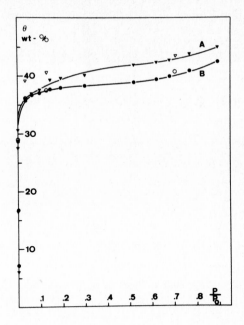

Fig. 1. Adsorption isotherms of $Fe(CO)_5$ on (A) HY- and (B) NaY-zeolite at 20°C [p_o = 29.4 mbar]. Full points : adsorption; open points : desorption.

pressures and remains almost constant up to $p/p_o \approx$ 0.5. At this partial pressure NaY and HY adsorb 39 and 42 mg of $Fe(CO)_5$ per 100 mg of dry zeolite, respectively.

Desorption is reversible down to ca. p/p_o = 0.1. After degassing for 15 hrs at 10^{-5} mbar, $Fe(CO)_5$ loadings of 29 and 27 wt % are obtained for NaY and HY, respectively.

The adsorption behaviour can be explained in terms of nearly ideal micropore adsorption (ref. 13), the micropores being the supercages of the faujasite. The amount $Fe(CO)_5$ adsorbed before capillary condensation occurs corresponds to 25 molecules/U.C. or 3.1 molecules per supercage. If an effective radius of 0.30 nm is assumed for the complex, the geometry of the supercages allows a maximum adsorption of three molecules per supercage.

This good agreement with the experimental results confirms the picture of completely filled supercages.

2. Infrared study of $Fe(CO)_5$ adsorbed on zeolite Y

The $Fe(CO)_5$ saturated zeolite wafers show no measurable transmission in the CO-stretching region. The IR spectra reported in Fig. 2 correspond therefore to samples loaded with about 10 % of the capacity obtained at saturation. The carbonyl vibrations show a rather similar pattern for both NaY and HY,

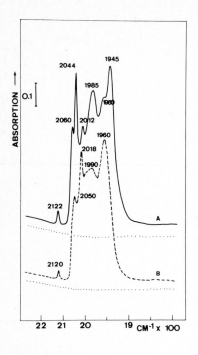

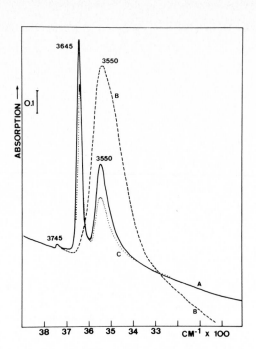

Fig. 2. (left) IR spectra of Fe(CO)$_5$/zeolite adducts at 20°C (saturated for 10 %). A : Fe(CO)$_5$/NaY; B : Fe(CO)$_5$/HY; dotted lines : zeolites degassed at 450°C.

Fig. 3. (right) OH spectrum of saturated Fe(CO)$_5$/HY adduct decomposed in 600 mbar He. A : zeolite degassed at 450°C; B : HY saturated with Fe(CO)$_5$ at 20°C; C : sample B after heating at 150°C for 45 min.

respectively (Table 1[a]). Compared to the HY-adduct, Fe(CO)$_5$ adsorbed on NaY exhibits two additional bands at 2044 and 1945 cm^{-1}, all other bands only slightly being changed.

The carbonyl bands cannot be assigned definitely to particular species inside the zeolite cages. The assignment of the carbonylbands to monosubstituted species as ZOH---Fe(CO)$_4$ (ref. 11,12) seems to be somewhat arbitrary because of the fortuitous agreement of some bands with those of complexes such as Fe(CO)$_4$P(CH$_3$)$_3$.

By ^{13}C-nmr line-width broadening, a restricted mobility of the Fe(CO)$_5$ adsorbed in HY was found (10). This is quite reasonable since the adsorption experiments indicate complete filling of the supercages. The carbonyl bands of crystalline Fe(CO)$_5$ and of the Fe(CO)$_5$/zeolite adducts (Table 1) show fairly good agreement, although the intensities are different. This also is in line with the restricted mobility of the complex in the supercages.

In particular, the appearence of the sharp 2122 cm^{-1} band can be explained by decreased site symmetry. Indeed, the intensity of the ν_1 mode increases on

TABLE 1

IR frequencies of $Fe(CO)_5$ and its adducts (CO-stretching region).

Adduct	Frequency/cm^{-1}							Ref.
$Fe(CO)_5$/NaY	2122w	2060s	2044s	2012s	1985s	1960sh	1945s	a
$Fe(CO)_5$/HY	2120w	2050s		2018s	1990b	1960s		a
$Fe(CO)_5$/HY	2112mw	2040s	2030sh	2010s	1985sh	1950ms		12
$Fe(CO)_5$ solid (-173°C)	2115(ν_1)	2033(ν_2)	2017sh	2003(ν_6)	1980(ν_{10})	1956/48(^{13}CO)		14
$Fe(CO)_5$ liquid (25°C)		[broad]		2002(ν_6)	1979(ν_{10})			15
$Fe(CO)_5$ gas (25°C)			2034vvs(ν_6)		2014vvs(ν_{10})			16

^{a}This work.

going from the gas phase to crystalline state (ref. 14).

The IR spectrum of liquid $Fe(CO)_5$ exhibits a broad and poorly structurated CO stretching band around 2000 cm^{-1} (Table 1). This also is in contrast to the observed well structurated CO-bands of the $Fe(CO)_5$/zeolite adducts.

In both the NaY and the HY adduct no ν_{CO}-bridging are shown, i.e., at 20°C no clusters with bridging CO ligands such as $Fe_2(CO)_9$ are generated.

The interaction of $Fe(CO)_5$ with the OH-bands of HY is illustrated in Fig. 3. Only the supercage hydroxyl groups (3645 cm^{-1}) disappear completely upon adsorption of the carbonyl. The band at 3550 cm^{-1} broadens and increases in intensity. This can only be explained by the formation of an hydrogen bond of moderate strength between the complex and the supercage OH-groups.

3. Desorption of $Fe(CO)_5$ from Y zeolite

Heating the $Fe(CO)_5$ loaded zeolite wafers in a vacuum of 10^{-4} mbar results in a proportional decrease of intensity of all carbonyl bands, while no new bands appear (Fig. 4).

When the similar experiments are carried out on larger amounts of sample (100 mg), iron losses between 20 and 50 % with respect to carbonyl saturation are determined gravimetrically.

These observations must be explained by desorption of the iron complex, which also may be the reason for the different iron loadings obtained by other authors (ref. 10-12) after thermal decomposition in vacuum, with values ranging from 1 to 8 wt % Fe. The proportional decrease of all carbonyl bands (Fig. 4) also indicates that one single species is adsorbed in the zeolite cage, which seems to be the intact $Fe(CO)_5$.

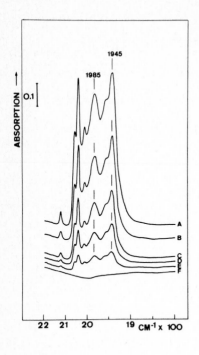

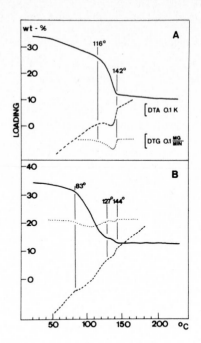

Fig. 4. (left) IR spectra of the desorption process of the 10 % saturated
Fe(CO)$_5$/Na adduct upon heating in vacuum.
A : initial loading at 20°C; B : heating at 40°C for 45 min; C : at 60°C for
30 min; D : at 65°C for 20 min; E : at 65°C for 40 min; F : at 65°C for 100 min.

Fig. 5. (right) Thermogramm of the decarbonylation of Fe(CO)$_5$/zeolite adducts
in He flow. (Heating rate : 1°C/min to 200°C).
A : Fe(CO)$_5$/NaY (saturated), 8.90 mg dry wt; B : Fe(CO)$_5$/HY (saturated, 9.65 mg
dry wt. Full line = TG curve; dashed line = DTA; dotted line = DTG.

4. Decarbonylation of Fe(CO)$_5$/zeolite adducts by thermoanalysis

   When in a thermobalance different amounts of zeolite are loaded with carbonyl vapor, the saturation loadings correspond to the adsorption isotherms
(38 wt %). Samples are heated up to 200°C in a He stream at rates from 0.2 to
2°C/min.

   a. Decomposition of Fe(CO)$_5$/NaY. Irrespective of the amount of sample and
the heating rate, three distinct regions are found with respect to thermal
behaviour. Two zones of slow weight loss are separated by a fast decrease in
sample weight (Fig. 5A). The latter is accompanied by an endothermic DTA effect.
It is striking, that the DTA effect always starts, when the sample has lost
15 wt % of its loading. The begin of the third zone is defined by the end of
the DTA effect and always occurs when 26 wt % of loading are lost. Around 200°C,
the sample weight becomes stable and corresponds to a loading of 10.5 %.

   Begin- and end-temperature of the DTA effect are strongly dependent on the

heating rate. Extrapolation of these temperatures to zero heating rate
(ref. 17) indicates, that the same decomposition can be performed isothermally
in the temperature region between 70 and 90°C. An isothermal experiment at 90°C
after 11 hrs showed the break in the weight curve. Since the sample weight
does not change from 200 to 400°C, the adsorbate present is considered to be
metallic iron. Compared to the original carbonyl loading losses of iron must
be smaller than 0.5 wt %.

This is in contrast to the results of vacuum decomposition and can be explained by an efficient hindering of the carbonyl diffusion at high pressures
of inert gas.

The thermoanalytical results allow to depict carbonyl decomposition on NaY
as follows :

$$NaY/Fe(CO)_5 \xrightarrow{\text{slow}} Fe(CO)_2 \underset{\text{DTA}}{\xrightarrow{\text{fast}}} Fe(CO)_{1/4} \xrightarrow{\text{slow}} Fe$$

The agreement between the measured weight loss and the one calculated according
to this stoichiometry lies within 5 %.

b. Decomposition of $Fe(CO)_5$/HY. The thermogramm of the HY/$Fe(CO)_5$ adduct
is distinctly different from the one of the NaY adduct (Fig. 5B). First, the
fast decomposition as indicated by the start of the DTA effect always appears
at lower temperatures (83°C). Second, two weak, but reproducable endotherm DTA
effects are observed instead of one, which correlate with the DTG minima. Above
144°C (the endpoint of the second DTA effect) no further weight loss occurs.
Similar considerations as with NaY lead to the following stages of decomposition :

$$HY/Fe(CO)_5 \xrightarrow{\text{slow}} Fe(CO)_4 \underset{\text{1. DTA}}{\xrightarrow{\text{fast}}} Fe(CO) \underset{\text{2. DTA}}{\xrightarrow{\text{fast}}} Fe$$

Again, excellent agreement between calculated and measured values is obtained.

In previous work decomposition of $Fe(CO)_5$/HY in vacuum was reported to start
at 25°C and to be complete at 200°C (ref. 12). The formation of $Fe(CO)$/HY is
postulated in vacuo at 70°C (ref. 10). In these studies no reaction times were
reported.

From our results it is clear that the temperature for complete decomposition
is below 90°C and that decomposition is a very slow reaction.

5. In situ investigation of the Fe(CO)$_5$/zeolite decomposition by IR spectroscopy

Decomposition of the carbonyl adducts in a He atmosphere leads to results quite distinct from those of vacuum heating. For the Fe(CO)$_5$/NaY this is shown in Fig. 6. After heating a NaY sample (saturated with Fe(CO)$_5$) at 150°C for 10 min, a broad band around 1940 cm^{-1} with a shoulder at 1860 cm^{-1} is generated, replacing the 1985, 1960 and 1945 cm^{-1} bands of the original adduct (Fig. 6D).

The original high frequency bands, in particular the 2044 cm^{-1} vibration, strongly decrease in intensity after prolonged heating at 150°C, leaving a shoulder at the 2005 cm^{-1} band and a weak band at 2070 (Fig. 6E).

The broad low frequency band changes into a vibration at 1900 cm^{-1}, which is the last band to survive by further heating to 200°C. Decomposition is complete after some hours at 200°C. No bands below 1800 cm^{-1} are observed during decomposition.

For the Fe(CO)$_5$/HY adduct the decomposition is given in Fig. 7. Thermal treatment at 130°C of a saturated HY-adduct first leads to the generation of a new band around 1880 cm^{-1} (Fig. 7D), whereas after 7 min. a relative decrease of the low frequency bands at 1960 and 1880 cm^{-1} occurs (Fig. 7E). Prolonged heating at this temperature causes rapid decomposition. After 12 min. only three weak vibrations at 2065, 2030 and 2000 cm^{-1} are left. Decomposition is completed after 45 min. at 150°C, and at the same time the original OH-bands are restored to about 75 % of their initial intensity (Fig. 3C).

In both zeolites bands below 1900 cm^{-1} are formed during the early decomposition, providing some evidence for bridged CO (ref. 18). They may be associated with Fe(CO)$_x$ species formed during the first phase of thermal decomposition. Bands at 1760 and 1790 cm^{-1}, which seemed to be characteristic for the Fe$_3$(CO)$_{12}$/HY adduct (ref. 12), have never been observed in the present case. The following indications exist for the intermediate formation of Fe$_3$(CO)$_{12}$:

- on the average 3 Fe(CO)$_5$ are adsorbed per supercage;
- the average stoichiometry after the first reaction step with HY is Fe(CO)$_4$;
- bands around 1880 cm^{-1} which is in the region for bridged CO are also observed for Fe$_3$(CO)$_{12}$ in Ar matrix (ref. 19) or in KBr pellets and in solution (ref. 12). The previously observed bands at 1760/90 cm^{-1} alternatively can be assigned to surface carbonate (ref. 20,21).

With Fe(CO)$_5$/NaY, the broad band around 1895 cm^{-1} is the dominant species before reaction is complete.

In general, LFe(CO)$_x$ species exhibit a decrease of CO stretching frequency with decreasing x, if L is a set of the corresponding number of electron donor ligands or an inert matrix (ref. 15,22-24). The present IR results for Fe(CO)$_5$/NaY can be understood in the same way, indicating a low CO coordination

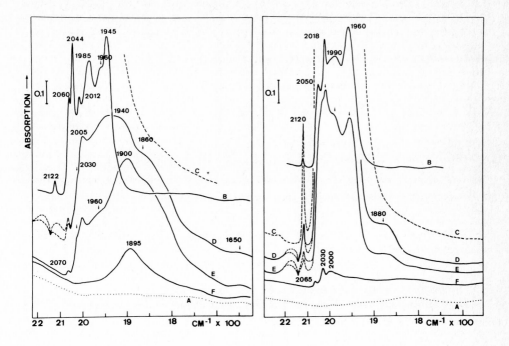

Fig. 6. (left) IR spectra of the decarbonylation of saturated Fe(CO)$_5$/NaY in He atmosphere.
A : NaY zeolite degassed at 450°C; B : Fe(CO)$_5$/NaY adduct (10 % saturated) at 20°C; C : Fe(CO)$_5$/NaY adduct (saturated) heated in 0.6 bar He at 100°C for 15 min.; D : sample C heated at 150°C for 10 min.; E : sample C heated at 150°C for 30 min., He pumped off; F : sample E heated at 200°C for 70 min. in 0.6 bar He; A : sample E heated at 200°C for 5 h.

Fig. 7. (right) IR spectra of the decarbonylation of saturated Fe(CO)$_5$/HY in He atmosphere.
A : HY zeolite degassed at 450°C; B : Fe(CO)$_5$/HY adduct (7 % saturated) at 20°C; C : Fe(CO)$_5$/HY adduct (saturated) at 20°C; D : sample C heated up to 130°C in 0.6 bar He; E : sample D heated at 130°C for 7 min.; F : sample D heated at 130°C for 12 min., He pumped off; A : sample D heated at 150°C for 45 min. in 0.6 bar He.

number of the last generated intermediates.

The lower thermal stability of the HY adduct is explained by weaker π-back-bonding towards the CO ligands due to the increased electron deficiency of the iron clusters. The effect of acidity on the metal-CO bond was also observed with PdHY zeolites (ref. 25).

The HY-hydroxyl groups of the adduct are only partially restored after decomposition, indicating the consumption of protons according to

$$Fe(CO)_5 + 2H^+ \longrightarrow H_2 + Fe(II) + 5CO \qquad (ref. 12)$$

The portion of oxidized iron, taking into account an initial proton content of 50 H^+/U.C., may therefore be estimated to be ca. 25 % of the iron loading.

CONCLUSION

The present work shows that the iron carbonyl at 20°C is strongly adsorbed until the supercages of the zeolites are saturated with three molecules on the average. The $Fe(CO)_5$ molecule remains intact on adsorption and is encaged in the zeolite with restricted mobility.

In HY, a hydrogen bond of moderate strength is formed with the supercage hydroxyls, which are completely involved in this process. Thermal decomposition in helium of the adducts leads to distinct CO-Fe fragments of different composition. Finally a reproducable iron loading of 10.5 ± 0.5 wt % is obtained. New evidence is found for the intermediate generation of $Fe_3(CO)_{12}$ during the decomposition in HY.

From the partly reversible hydroxyl interaction with the complex in HY, it is estimated that about one quart of the iron is oxidized during decomposition. Thermal decomposition is a slow reaction which goes to completion already between 70 and 90°C. It proceeds faster in case of HY due to acidic destabilization of the Fe-CO bond.

Work is in progress to determine the parameters influencing the particle size and catalytic properties of these zeolite supported iron clusters.

ACKNOWLEDGEMENTS

The technical assistance of Hugo Leeman is highly appreciated. One of us (T.B.) is indebted to the DAAD (Deutscher Akademischer Austauschdienst) and the belgian "Ministerie van Nationale Opvoeding en Nederlandse Cultuur" for a grant. P.A.J. acknowledges permanent research position as "Onderzoeksleider" from the Belgian Science Foundation (N.F.W.O.-F.N.R.S.). Financial support from the same institution and from the belgian government (Geconcerteerde Actie Catalyse, Diensten Wetenschapsbeleid) is gratefully acknowledged.

REFERENCES

1   Kh.M. Minachev, Y.I. Isakov in "Zeolite Chemistry and Catalysis" (J.A. Rabo, ed.) A.C.S., Washington D.C., 1976, p. 552.
2   P.A. Jacobs, "Carboniogenic Activity of Zeolites", Elsevier, Amsterdam, 1977.
3   Y.-Y. Huang and J.R. Anderson, J. Catal., 40 (1975) 143.
4   R.L. Garten, W.N. Delgass and M.J. Boudart, J. Catal., 18 (1970) 90.
5   F. Schmidt, W. Gunsser and A. Knappwost, Z. Nat. Forsch., 30a (1975) 1627.
6   F. Schmidt, W. Gunsser and J. Adolph, A.C.S. Symp. Ser., 40 (1977) 291.
7   W. Gunsser, J. Adolph and F. Schmidt, J. Magn. Magn. Mater., (1980) 1115.
8   J.B. Lee, J. Catal., 68 (1980) 27.
9   A. Terenin and L.M. Roev, Spectrochim. Acta, 15 (1959) 946.
10 J.B. Nagy, M. Van Eenoo and E.G. Derouane, J. Catal., 58 (1979) 230.

11 D. Ballivet-Tkatchenko, G. Coudurier, H. Mozzanega and I. Tkatchenko in Fundament. Res. Homog. Catal. (Tsutsui ed.) New York, 1979, p. 257.

12 D. Ballivet-Tkatchenko and G. Coudurier, Inorg. Chem., 18 (1979) 558.

13 M.M. Dubinin, in Progr. Surf. Membr. Sci., D.A. Cadenhead et al., ed., vol. 9 (1975) 1-69.

14 R. Cataliotti, A. Foffani and L. Marchetti, Inorg. Chem., 10 (1971) 1594.

15 M. Bigorgne, J. Organomet. Chem., 24 (1970) 211.

16 W.F. Edgell, W.E. Wilson and R. Summitt, Spectrochimica Acta, 19 (1963) 863.

17 S. Tanaka, Bull. Chem. Soc. Japan, 38 (1965) 795.

18 L.H. Little, "Infrared Spectra of Adsorbed Species", A.P., London, 1966, pp. 51.

19 M. Poliakoff and J.J. Turner, J.C.S. Chem. Comm., (1970) 1008.

20 P.A. Jacobs, F.H. Van Cauwelaert, E.F. Vansant and J.B. Uytterhoeven, J.C.S. Faraday I, 69 (1973) 1056.

21- P.A. Jacobs, F.H. Van Cauwelaert and E.F. Vansant, J.C.S. Faraday I, 69 (1973) 2130.

22 B.F.G. Johnson, J. Lewis and M.V. Twigg, J.C.S. Dalton, (1974) 241.

23 M. Poliakoff and J.J. Turner, J.C.S. Dalton, (1974) 2276.

24 M. Poliakoff, J.C.S. Dalton, (1974) 210.

25 P. Gallezot, Catal. Rev.-Sci. Eng., 20(1) (1979) 121.

STABILIZATION AND CHARACTERIZATION OF METAL AGGREGATES IN ZEOLITES. CATALYTIC PROPERTIES IN CO + H$_2$ CONVERSION.

D. BALLIVET-TKATCHENKO, G. COUDURIER and NGUYEN DUC CHAU

Institut de Recherches sur la Catalyse - C.N.R.S. - 2, avenue Albert Einstein 69626 Villeurbanne Cédex (France).

ABSTRACT

Fe$_3$(CO)$_{12}$ and Co$_2$(CO)$_8$-NaY adducts were prepared as catalyst precursors in the hydrocondensation of carbon monoxide. The characterization of these systems by IR, UV-visible and TPD techniques shows that the decarbonylation process occurs stepwise upon thermal treatment. Dismutation of the starting complexes appears as CO is evolved leaving cationic and anionic complexes in the zeolite. At the final stage of decarbonylation metallic iron particles and a mixture of cobalt species in high and very low oxidation states are formed. The presence of the metal inside the zeolite cavities shortens the chain-length of the hydro-carbons produced in CO + H$_2$ conversion.

INTRODUCTION

One interest in preparing metallic aggregates of iron and cobalt in zeolites is the study of their catalytic properties in the hydrocondensation of carbon monoxide (ref.1). Curiously the zeolite support has only received a limited attention in this area. In fact, this is due to the preparation method of metal-zeolite systems which is conventionally based on ion-exchange and further reducing treatments. Therefore, in the case of iron and cobalt, as it is very hard to reduce the corresponding ions to the metallic state (refs.2,3), CO + H$_2$ conversion is not observed. We have developed another preparation method based on the adsorption of volatile carbonyl compounds in the zeolite cavities. This method is more flexible because it allows to prepare samples without any protons, and in this case CO + H$_2$ conversion is observed (refs.4,5,6). This is a particularly interesting situation because the zeolite can play an important role during the catalytic process due to its physico-chemical properties (ref.7).

Fe(CO)$_5$, Fe$_2$(CO)$_9$, Fe$_3$(CO)$_{12}$, Co$_2$(CO)$_8$, Ru$_3$(CO)$_{12}$-Y zeolite adducts have been prepared (ref.5). The catalytic activity mainly depends upon the presence of protons according to the redox behavior reported in equation (1) :

$$2 \; M(0) + 2 \, n \, H^+ \rightleftharpoons 2 \, M^{n+} + nH_2 \qquad (1)$$

With iron and cobalt, the oxidation reaction is favored so that one has to start
with a non-acidic zeolite in order to prepare catalyst precursors.

In this paper, we essentially describe the characterization of $Fe_3(CO)_{12}$ or
$Co_2(CO)_8$-NaY zeolite adducts which are catalysts in $CO + H_2$ conversion. The
interactions of the carbonyl complexes $Fe_3(CO)_{12}$ and $Co_2(CO)_8$ with the NaY
zeolite were studied by three different techniques : UV-visible, infrared
spectroscopies and temperature-programmed desorption (TPD). By following the
transformation of the IR carbonyl bands of the clusters upon adsorption and
subsequent thermal treatment under vacuum, we have been able to characterize
a set of subcarbonyl species which finally lead to iron and cobalt particles
after decarbonylation. The TPD technique was also used to support the IR
interpretations by determining the nature of the gas evolved upon thermal
treatment and to precise if gas evolution is a monotoneous or a stepwise process.
The $Fe_3(CO)_{12}$ and $Co_2(CO)_8$-NaY adducts catalyze the Fischer-Tropsch reaction.
The selectivity of this reaction will be briefly mentionned in relation to the
location and size of the metal particles. Comparison with the $Ru_3(CO)_{12}$-NaY
system will also be made in order to present some trends in the role of the
zeolite.

EXPERIMENTAL

IR spectra were recorded at 25°C on a Perkin Elmer 580 spectrophotometer with
a resolution of 2.5 to 3.7 cm^{-1}. Diffuse reflection spectra were recorded in
the 220-2500 nm range with an Optica Milano CF_4Ni spectrophotometer using a
differential method with MgO or NaY zeolite as reference.

Simultaneous UV-visible and IR studies were conducted using a special
cell already described (ref.8). The zeolite wafers for IR studies were obtained
by compressing 6 to 15 mg of zeolite into 18 mm diameter disc at 0.4 $Ton.cm^{-2}$.
Disc of 22 mm diameter and 300 to 500 mg weight were prepared for UV measurements.

After a pretreatment of the zeolite for 15 h in oxygen and 3 h in vacuo
(10^{-5} Torr) at 350°C, a break seal device containing under vacuum the carbonyl
complex is connected to the cell, and the adsorption takes place at 25°C. The
amount of carbonyl compounds anchored is determined by chemical analysis.

For the TPD technique, 5-30 mg of $Fe_3(CO)_{12}$ or $Co_2(CO)_8$-NaY are transferred
under inert atmosphere in a reactor connected to a mass spectrometer (Riber
Quadrupole QS 200). The experiments are performed under vacuum by heating the
sample up to 600°C at a 4°C/min rate. Continuous analysis by mass spectrometry
of the gas evolved permits to get the TPD curves according to the procedure
developed by Massardier and Tran Manh Tri (ref.9). The samples are prepared by
sublimation of the carbonyl complexes in a similar device than that described
for the IR experiments.

The catalytic runs are performed in a 300 ml batch reactor under 20 bars initial pressure at 25°C during 17 hours with a CO/H_2 ratio of 1/3 (ref.7).

RESULTS AND DISCUSSION

$Fe_3(CO)_{12}$-NaY. Adsorption of $Fe_3(CO)_{12}$ and decomposition under vacuum.

IR study. Fig. 1 reports the characteristic IR bands of the carbonyl species obtained upon $Fe_3(CO)_{12}$ adsorption at room temperature and after subsequent thermal treatment under vacuum.

Adsorption of $Fe_3(CO)_{12}$ is accompanied by a color change to green (visible absorption : λ_{max} = 630 nm). No gas evolution is observed, and in the low IR frequency region the zeolite framework vibrations are not perturbed. The mode of interaction between $Fe_3(CO)_{12}$ and NaY can be depicted from the comparison between CO frequencies of $Fe_3(CO)_{12}$ in the different environments already known. For example, Table 1 reports the νCO and $\delta FeCO$ frequencies for $Fe_3(CO)_{12}$ (i) in n-hexane, (ii) in a HY matrix and (iii) in the NaY matrix. The IR spectra are very similar (number of bands, relative intensities and frequencies).

TABLE 1

Carbonyl frequencies for $Fe_3(CO)_{12}$ and HY, NaY adducts

$Fe_3(CO)_{12}$ [a] n-hexane	$Fe_3(CO)_{12}$-HY [a]	$Fe_3(CO)_{12}$-NaY [b]	Assignement
2103 (w)	2112 (w)	2122 (w)	
2046 (vs)	2056 (vs)	2061, 2046 (vs)	Terminal νCO
2023 (mb)	2030 (m)	2018 (m)	
2013 (sh)			
1867 (w)	1795 (w)	1805 (w)	Bridging νCO
1838 (m)	1760 (m)	1770 (m)	
639 (w, sh)	640 (m)	645 (m)	δ FeCO
594 (ms)	615 (sh)	625 (sh)	
570 (ms)			

(a) ref.4
(b) additional weak bands at 1988, 1962 (sh) and 1947 cm^{-1}.

Therefore we propose that the NaY behaves as the HY-zeolite (ref.4) like a solvent in which the $Fe_3(CO)_{12}$ has terminal and bridging carbonyl ligands. The bridging carbonyls and the zeolite cations (Na^+ or H^+) interact via an acid-base adduct which depicts the way of anchoring of this type of complex : $>C=O--Na^+(H^+)$.

Thermal treatment in vacuo at 60°C is accompanied by evolution of small quantity of gas and the IR spectrum changes (Fig. 1c). New intense bands at 2056 (m), 1988 (s), 1962 (sh), 1947 (vs) and 1760 (w) cm^{-1} are now present. Therefore other species is (are) preferentially formed at this stage. We have found that

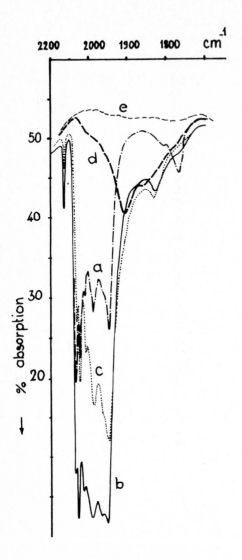

Fig. 1. IR spectra of
Fe$_3$(CO)$_{12}$-NaY (3.5 % Fe) :
(a) beginning of adsorption
of Fe$_3$(CO)$_{12}$ at 25°C,
(b) final stage of adsorption
at 25°C, (c) thermal treatment
at 60°C, (d) at 150°C,
(e) at 200°C.

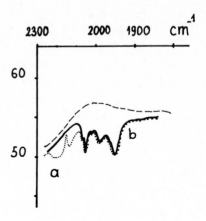

Fig. 2. IR spectra of
adsorbed CO (20 torr) on the
Fe$_3$(CO)$_{12}$-NaY sample pretreated at 200°C : (a) in the
presence of CO, (b) after
evacuation, at 25°C.

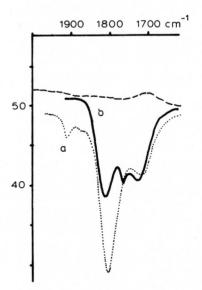

Fig. 3. IR spectra of
adsorbed NO (20 torr) on the
Fe$_3$(CO)$_{12}$-NaY sample pretreated at 200°C : (a) in the
presence of NO, (b) after
evacuation, at 25°C.

the same IR spectrum is obtained when a small amount of water (H_2O or D_2O, vapor pressure 4 torr) is admitted into the IR cell after adsorption of $Fe_3(CO)_{12}$ at 25°C and a further heating at 60°C of the hydrated sample. In order to know what kind of complex can be formed at 60°C we have examined in some detail the IR spectra of the hydrated sample at 25°C, after the thermal treatment at 60°C, together with those reported in the literature for species formed from $Fe_3(CO)_{12}$ in the presence of water either under basic or acidic conditions. The following equations reports the reactivity of $Fe_3(CO)_{12}$ in basic medium (ref.10) :

$$Fe_3(CO)_{12} + 2\ OH^- \longrightarrow |Fe_3(CO)_{11}|^{2-} + CO_2 + H_2O \qquad (2)$$

$$|Fe_3(CO)_{11}|^{2-} + H_2O \rightleftharpoons |HFe_3(CO)_{11}|^- + OH^- \qquad (3)$$

According to the basicity of the solution, a monoanionic and a dianonic carbonyl complexes are formed. In an acidic solution or with the HY zeolite, $Fe_3(CO)_{12}$ is oxidized into Fe^{2+} (Fe^{3+}) species (ref.4).

The IR spectra of $Fe_3(CO)_{12}$-NaY (this work) and -HY adducts (ref.4) are very different upon thermal treatment whereas those of $|Fe_3(CO)_{11}|^{2-}$ and $|HFe_3(CO)_{11}|^-$ are similar to the NaY systems (Table 2). Especially the more intense bands

TABLE 2

Carbonyl stretching frequencies for $|HFe_3(CO)_{11}|^-$, $|Fe_3(CO)_{11}|^{2-}$ and $Fe_3(CO)_{12}$-NaY-H_2O adducts

$\|HFe_3(CO)_{11}\|^{-(a)}$ Et_4N^+, C_6H_6	$\|HFe_3(CO)_{11}\|^{-(a)}$ Et_3NH^+, CH_3CN	$\|Fe_3(CO)_{11}\|^{2-(b)}$	$Fe_3(CO)_{12}$-NaY +H_2O at 25°C	$Fe_3(CO)_{12}$-NaY +H_2O at 60°C
	2073 (w)			
2008 (vs)	2004 (vs)		2010 (vs)	1988 (m)
2000 (vs)	1980 (s)		2000 (vs)	1962 (sh)
	1946 (m)			1947 (vs)
		1938 (s)		1920 (sh)
		1910 (ms)		
		1890 (sh)		
1639 (m)	1742 (m)	1670 (w)		1760 (w)

(a) ref.11, (b) ref.12.

arise at similar frequencies. Some discrepancies exist in the bridging carbonyl frequency region but on these ionic complexes these bands are highly sensitive to the natures of the solvent and of the counter-cation. For example, with increasing solvent polarity the bridging νCO shifts from 1639 to 1742 cm^{-1}. On the basis of this observation, comparison of IR frequencies reported in Table 2 shows that reactions (2) and (3) take place in the hydrated NaY zeolite in the following way :

$$Fe_3(CO)_{12} \xrightarrow[25°C]{NaY-H_2O} |HFe_3(CO)_{11}|^- \qquad \nu CO : 2010-2000 \text{ (vs) cm}^{-1} \qquad (4)$$

$$|HFe(CO)_{11}|^- \xrightarrow[60°C]{NaY-H_2O} |Fe_3(CO)_{11}|^{2-} \qquad \nu CO : 1947 \text{ (vs) cm}^{-1} \qquad (5)$$

Therefore water in a NaY zeolite exhibits a basic behavior probably because ion-exchange between Na^+ and H_3O^+ is possible (equation (6))

$$2 H_2O + Na^+ - Z^- \longrightarrow OH^- + Na^+ + H_3O^+ - Z^- \qquad (6)$$

Coming back to the IR spectrum of $Fe_3(CO)_{12}$-NaY treated at 60°C in the absence of water, it can be assigned to the presence of $|Fe_3(CO)_{11}|^{2-}$ as it corresponds also to that of $Fe_3(CO)_{12}$-NaY-H_2O heated at 60°C (equation (5)). The formation of this complex can occur according to equation (2) if we assume that non detectable trace amount of water is present. But CO_2 will also be evolved at the same time. No CO_2 is detected either by IR or by mass spectro-metry, no carbonato species are detected by IR. Hence the transformation of $Fe_3(CO)_{12}$ into $|Fe_3(CO)_{11}|^{2-}$ proceeds via another mechanism. It is known that dismutation reaction of $Fe_3(CO)_{12}$ also forms, in anhydrous solution, the dianion with concomitant CO evolution (ref.13) :

$$4 Fe_3(CO)_{12} \longrightarrow 3|Fe|^{2+} |Fe_3(CO)_{11}|^{2-} + 15 CO \qquad (7)$$

In our experiments, during the thermal treatment of the NaY adduct at 60°C only CO is evolved as evidenced by mass spectrometry. Therefore equation (7) would correspond to the reactivity of $Fe_3(CO)_{12}$ in the non-hydrated zeolite.

Upon further thermal treatment above 60°C, a rapid CO evolution is observed at 100°C, and new subcarbonyl species is (are) formed at 150°C with a broad νCO band centered at 1900 cm^{-1} (Fig. 1d). Adsorption of CO (4 torr) at 25°C on this sample restores a set of strong carbonyl bands centered at 2060 (m), 2040 (ms), 2010 (vs) and 1950 (m) cm^{-1}. Such bands (number and frequencies) strongly indicates that neutral zerovalent iron carbonyl complexes are reformed and that the species exhibiting the band at 1900 cm^{-1} is a highly unsaturated carbonyl complex. This 1900 cm^{-1} band totally disappears at 200°C leading to a completely decarbonylated sample (Fig. 1e). Parallely the sample turns from brown to grey. Readsorption of CO (20 torr) at 25°C restores a set of bands but of weaker intensities (Fig. 2, 2060, 2045, 2015, 1975 and 1940 cm^{-1}). Therefore it appears that aggregation of iron could have taken place.

Adsorption of NO at 25°C (20 torr) on the 200°C pretreated sample gives νNO bands centered between 1805 and 1720 cm^{-1} (Fig. 3). These frequencies are lower than those reported for oxidized Fe complexes (refs.4,14) and lie in the same region than those reported for Fe(o) complexes (ref.15) and metallic ion (ref.16).

To conclude, $Fe_3(CO)_{12}$ is anchored in the NaY zeolite through zeolite cations-iron complex bridging carbonyls bonds. Total decarbonylation occurs upon thermal treatment at 200°C leaving metallic iron particles. Decarbonylation proceeds in stepwise manner during which $Fe_3(CO)_{12}$ undergoes reactions which are formally based on reversible dismutation process.

TPD study. The TPD experiment performed on $Fe_3(CO)_{12}$-NaY sample (2.7 % Fe) reveals a stepwise decarbonylation between 25° and 200°C with maxima at 70°, 100° and 175°C that nicely fit with the IR study (Fig. 4c). Additional thermo-desorption peaks are not observed up to 600°C, and mass spectrometry analysis shows that CO is the only gas evolved.

Fig. 4a reports the blank TPD curve with unsupported $Fe_3(CO)_{12}$. A single narrow peak occurs at 85°C ; no discrete subcarbonyl complexes are formed ; a metallic miror is observed after decarbonylation. Therefore the NaY zeolite matrix profoundly modifies the course of decarbonylation. Similar stabilizing effect has also been recently reported with alumina support whereas silica behaves differently (ref.17).

Other modifications in the TPD curves have been found upon iron content. Taking into account the size of $Fe_3(CO)_{12}$ (7.0 x 7.2 x 8.3 Å) and the diameter of the zeolite windows and cavities, $Fe_3(CO)_{12}$ can be located only in the super-cages ($\simeq$ 7.5 x 12.5 Å) after adsorption. Hence the theoretical maximum loading leads to a value of $\simeq$ 11 % Fe. Three representative samples were prepared : 0.97 % (low loading), 2.7 % (medium loading) and 9.7 % Fe (saturation loading). Fig. 4b,c,d respectively gives the different TPD curves. At low and medium iron loadings the curves are identical. Three CO peaks are present. The first peak, at 70°C, is of medium height and would correspond to departure of $\simeq$1.2 CO per Fe for the formation of $|Fe_3(CO)_{11}|^{2-}$ (equation (7)). The second peak at c.a. 100°C is the highest and corresponds to decomposition of $|Fe_3(CO)_{11}|^{2-}$ into a subcarbonyl complex which exhibits a νCO broad band at 1900 cm^{-1}. This unidentified carbonyl species evolves CO between 150° and 200°C in very small quantity : certainly much less than 1.2 CO per Fe if one roughly estimates the surface areas corresponding to the first and the third peaks. Hence this species is highly unsaturated with an iron framework of several atoms stabilized by very few carbonyl ligands. It is noteworthy that it readsorbs CO thus leading to some neutral carbonyl species (see IR section). The 9.7 % Fe sample is quite different (Fig. 4d). Only two very close narrow CO desorption peaks of approximately equal intensities occur at 85° and 95°C. It appears that the presence

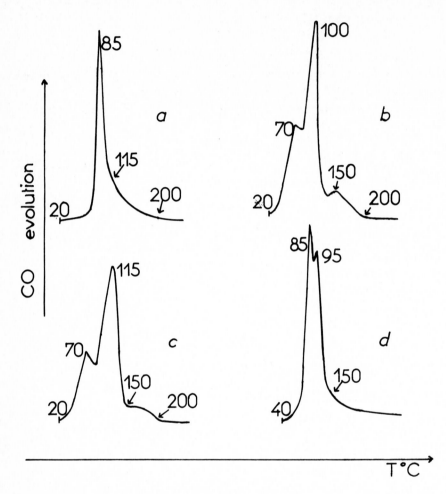

Fig. 4. TPD of : (a) unsupported Fe3(CO)12, (b) Fe3(CO)12-NaY (0.97 % Fe),
(c) Fe3(CO)12-NaY (2.7 % Fe), (d) Fe3(CO)12-NaY (9.7 % Fe).

of the zeolite does not drastically modify the TPD of $Fe_3(CO)_{12}$. After the
experiment a metallic miror is detected on the reactor walls. Hence, a certain
amount of $Fe_3(CO)_{12}$ is not located inside the zeolite.

$Co_2(CO)_8$-NaY. Adsorption of $Co_2(CO)_8$ and decomposition under vacuum.

Fig. 5 reports the characteristic IR bands of the carbonyl species obtained
upon $Co_2(CO)_8$ adsorption at room temperature and after subsequent thermal
treatment under vacuum. The spectra are more complex than those of $Fe_3(CO)_{12}$.

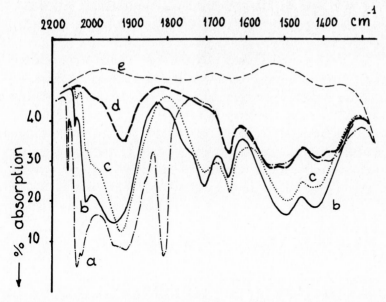

Fig. 5. IR spectra of $Co_2(CO)_8$-NaY (3.3. % Co) : (a) adsorption of $Co_2(CO)_8$ at 25°C, (b) thermal treatment at 50°C, (c) at 100°C, (d) at 150°C, (e) at 200°C.

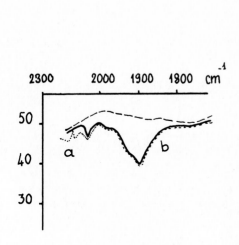

Fig. 6. IR spectra of adsorbed CO (20 torr) on the $Co_2(CO)_8$-NaY sample pretreated at 200°C : (a) in the presence of CO, (b) after evacuation, at 25°C.

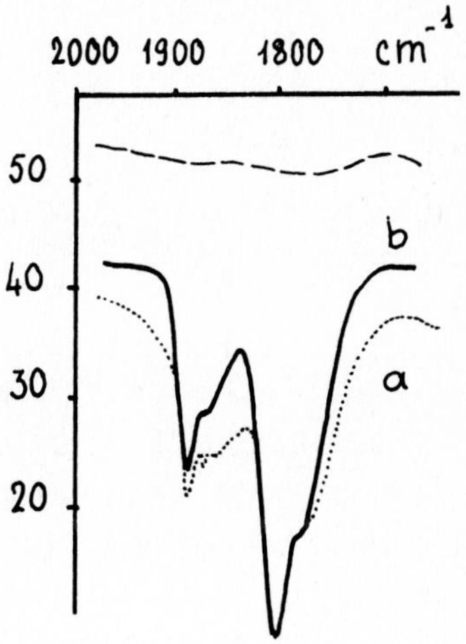

Fig. 7. IR spectra of adsorbed NO (20 torr) on the $Co_2(CO)_8$-NaY sample pretreated at 200°C : (a) in the presence of NO, (b) after evacuation, at 25°C.

The TPD curve is also more complex. Total decarbonylation occurs at 300°C instead of 200°C with $Fe_3(CO)_{12}$. Four peaks are detected one at very low temperature 40°C, a broad one at 130°C and two others at 200 and 250°C. A parallel study on other ion-exchanged Y zeolites (CaY, HY, LaY) has provided a much better understanding of the reactivity of $Co_2(CO)_8$ in a NaY matrix. Herein are presented preliminary results and interpretations which are based on the extrapolation of $Co_2(CO)_8$ chemistry in solution.

The IR spectrum obtained after adsorption at 25°C (Fig. 5a) does not corres-pond to the adsorbed formed of $Co_2(CO)_8$ with retention of its stereochemistry (Table 3). Comparison with IR data reported in the series of cobalt carbonyl complexes permits to discriminate three species. The set of bands at 2120 (w),

TABLE 3

Carbonyl frequencies for $Co_2(CO)_8$, $Co_4(CO)_{12}$ and selected ones for $Co_2(CO)_8$-NaY adduct.

$Co_2(CO)_8$ n-heptane	$Co_4(CO)_{12}$[a] n-hexane	$Co_2(CO)_8$-NaY	Assignement
2112 (vw)	2104 (w)	2122 (w)	
2069 (s)	2063 (vs)	2080 (vs)	
2055 (s)	2055 (vs)	2055 (s)	Terminal νCO
2045 (s)	2048 (sh)		
2030 (s)	2038 (w)		
2022 (s)	2028 (w)		
2001 (w)			
1991 (wb)			
1866 (m)	1867 (s)	1826 (s)	Bridging νCO
1857 (s)			
545 (mw)		542 (m)	δ CoCO
528 (m)			

(a) ref. 18.

2082 (vs), 2052 (s) cm^{-1} (terminal CO ligands) and 1810 (s) cm^{-1} (bridging CO) fits with those of unsupported $Co_4(CO)_{12}$ (Table 3) if one assumes that the down shift frequency (40 cm^{-1}) of the bridging carbonyls is due, as for $Fe_3(CO)_{12}$ (Table 1), to the acid-base adduct with Na^+ zeolite cations. The formation of $Co_4(CO)_{12}$ from $Co_2(CO)_8$ is a well-known process either under inert atmosphere or under vacuum (ref.19) :

$$2\ Co_2(CO)_8 \underset{+CO}{\overset{-CO}{\rightleftharpoons}} Co_4(CO)_{12} \tag{8}$$

The set of bands at 1945 (sh), 1942 (s) and 1912 (s) cm^{-1} together with the presence of a visible absorption band at 375 nm are characteristic of the anion $|Co(CO)_4|^-$ in a distorted tetrahedra arrangement (refs.19,20). Other two intense broad bands at 1500 and 1420 cm^{-1} lie in the region of carbonato species. Such carbonato species indicates that CO_2 is formed during the transformation of $Co_2(CO)_8$. In addition, weak bands at 3695 and 1640 cm^{-1} attributed to H_2O appear during $Co_2(CO)_8$ adsorption. Hence a secondary reaction can take place with water : free OH^- groups in the zeolite (equation (6)) attack the neutral carbonyl species. In basic solution, $Co_2(CO)_8$ and $Co_4(CO)_{12}$ react according to equations (9) and (10) (ref.10) :

$$11\ Co_2(CO)_8 + 32\ OH^- \rightarrow 2\ Co^{2+} + 20|Co(CO)_4|^- + 8\ CO_3^{2-} + 16\ H_2O \qquad (9)$$

$$11\ Co_4(CO)_{12} + 16\ OH^- \rightarrow 12\ Co^{2+} + 32|Co(CO)_4|^- + 4\ CO_3^{2-} + 8\ H_2O \qquad (10)$$

$|Co(CO)_4|^-$, Co^{2+} and CO_3^{2-} species are now present. On the zeolite system, $|Co(CO)_4|^-$ and CO_3^{2-} have been identified by IR. The Co^{2+} is also detected by UV-visible : bands at 525, 575 and 610 nm are relevant of the Co^{2+} species in a tetrahedral environment (ref.8).

Upon thermal treatment in vacuo many changes occur in the IR and UV-visible spectra. The IR bands attributed to $Co_4(CO)_{12}$ decrease at 50°C (Fig. 5b) and are totally absent at 100°C (Fig. 5c). New bands appear in the region 1740-1640 cm^{-1}. Between 100-150°C, the high frequencies, νCO, disappear, and a broad band at 1920 cm^{-1} remains (Fig. 5d) and progressively disappears above 150°C. At 200°C the IR spectrum exhibits no more CO bands. At this stage admission of CO (20 torr) at 25°C into the IR cell restores CO bands at 2060 (w), 1920 (sh) and 1900 (s) cm^{-1} (Fig. 6). CO adsorbed on metallic cobalt gives rise to bands centered at higher frequencies : 1970 cm^{-1} on Co film (ref.22), 2060-2020 cm^{-1} on Co-SiO$_2$ (ref.23) and 2069 (m), 1977 (s), 1968 (sh), 1955 (sh) cm^{-1} on Co^{2+}-Cd-A zeolite (ref.24). In our case, we propose that the cobalt is in a cluster arrangement bearing anionic charges. Further study is underway in order to clarify this point. The adsorption of NO (20 torr) at 25°C into the IR cell after total decarbonylation produces a complicated spectrum (Fig. 7) which in fact corresponds to two sets of two bands. The 1890 and 1808 cm^{-1} bands are relevant of a Co^{2+} nitrosyl complex in agreement with refs.25 and 26. The 1870 and 1775 cm^{-1} bands are relevant of a Co nitrosyl complex in lower oxidation state (ref.15).

To summarize, $Co_2(CO)_8$ is readily transformed in a NaY zeolite into $Co_4(CO)_{12}$ which is stabilized through acid-base interaction between the bridging carbonyl ligands and the Na^+ zeolite cations. Other species are formed either upon adsorption or upon thermal treatment. Among these species Co^{2+} and $|Co(CO)_4|^-$ have been characterized. Finally, decarbonylation at 200°C leaves Co^{2+} ions and other cobalt species in lower oxidation state.

Some trends in selectivity for CO + H_2 conversion

One might expect that a better control on the metallic particle size could improve the selectivity in obtaining a narrower range of products (ref.1). Some results have already been obtained in that direction (refs.5,7,27,28). A further investigation of our systems, $Ru_3(CO)_{12}$-Y and $Co_2(CO)_8$-NaY, shows that the situation is more complicated.

CO + H_2 conversion is observed if the catalyst precursor contains the metal in a reducible form : cluster framework of low oxidation state or metallic particles. Saturated hydrocarbons are the main products.

Three Ru samples were selected according to their electron microscopy pictures and tested in CO + H_2 conversion at 200°C. One sample contains exclusively metallic particles located in the zeolite cavities (15-20 Å diameter). 95 % of the carbon-containing products lie in the range C_1-C_9 with a maximum at C_4-C_5. Another sample contains metallic particles exclusively outside the zeolite (20-80 Å diameter) with ill-defined shapes. 33 % of the carbon-containing products lie in the range C_1-C_9 and waxes stick on the zeolite are observed. The third sample contains metallic particles exclusively outside the zeolite (hexagonal plates ≈500-1000 Å diameter). 100 % of methane is now observed. On the cobalt systems, CO + H_2 conversion is only significant at 250°C. A sample containing exclusively metallic particles inside the zeolite (30-40 Å in diameter) leads to 95 % of hydrocarbons in the C_1-C_{10} range with 71 % selectivity in CH_4, whereas a sample containing particles inside and outside the zeolite (≈ 100 Å diameter) leads to 91 % of hydrocarbons in the C_1-C_{12} range with 40 % selectivity in CH_4. It appears from these experiments that the location of the metal inside the zeolite induces a shortening of hydrocarbon chain-length for ruthenium and cobalt but with this latter it orientates the reaction towards methanation reaction.

CONCLUSION

The preparation of $Fe_3(CO)_{12}$ and $Co_2(CO)_8$-NaY zeolites are valuable candidates to catalyze CO + H_2 conversion. When this work was undertaken, we thought that a thermal treatment would lead straight forward to metallic particles. In fact the work described here points out that the zeolite orientates the decarbonylation via a more complicated route. The zeolite has two main properties : solvent and reactant. A great similarity has been found between the chemistries of the carbonyl complexes in solution and in the zeolite.

ACKNOWLEDGEMENTS

We thank Dr J. Massardier for its assistance and for the use of its apparatus for the TPD experiments, Mrs I. Mutin for the electron micrographs and Dr I. Tkatchenko for encouragements.

REFERENCES

1 M.E. Dry, in J.R. Anderson and M. Boudart (Ed.), Catalysis : Science and Technology, Vol. 1, Springer Verlag, New-York, 1981, Ch. 4, p. 159.
2 W.N. Delgass, R.L. Garten and M. Boudart, J. Phys. Chem., 73(1969)2970-2979.
3 R.L. Garten, W.N. Delgass and M. Boudart, J. Catal. 18 (1970) 90-107.
4 D. Ballivet-Tkatchenko and G. Coudurier, Inorg. Chem., 18 (1979) 558-564.
5 D. Ballivet-Tkatchenko, Nguyen Duc Chau, H. Mozzanega, M.C. Roux and I. Tkatchenko, Catalytic Activation of Carbon Monoxide, ACS Symp. Ser. n° 152, 1981, pp. 187-201.
6 D. Ballivet-Tkatchenko, G. Coudurier and H. Mozzanega, Catalysis by Zeolites, Elsevier, Amsterdam, 1980, pp. 309-317.
7 D. Ballivet-Tkatchenko and I. Tkatchenko, J. Molecular Catal., 13 (1981) 1-10.
8 H. Praliaud and G. Coudurier, J. Chem. Soc. Faraday I, 75 (1979) 2601-2616.
9 J. Massardier and Tran Manh Tri, private communication.
10 H. Behrens, J. Organomet. Chem., 94 (1975) 139-159 and references herein.
11 J.R. Wilkinson and L.J. Todd, J. Organomet. Chem., 118 (1976) 199-204.
12 F. Kip-Kwai Lo, G. Longoni, P. Chini, L.D. Lower and L.F. Dahl, J. Am. Chem. Soc., 102 (1980) 7691-7701.
13 K. Farmery, M. Kilner, R. Greatrex and N.N. Greenwood, J. Chem. Soc. A, (1969) 2339-2345.
14 J.W. Jermyn, T.J. Johnson, E.F. Vansant and J.H. Lunsford, J. Phys. Chem. 77 (1973) 2964-2969.
15 N.G. Connelly, Inorg. Chim. Acta, 6 (1972) 47-89.
16 G. Blyholder and M.C. Allen, J. Phys. Chem., 69 (1965) 3998-4004.
17 A. Brenner and D.A. Hucul, Inorg. Chem., 18 (1979) 2836-2840.
18 G. Bor, Spectrochim. Acta, 19 (1963) 1209.
19 G. Bor and U.K. Dietler, J. Organomet. Chem., 191 (1980) 295-302 and references herein.
20 W.F. Edgell and A. Barbetta, J. Am. Chem. Soc., 96 (1974) 415-423.
21 C. Schramm and J.I. Zink, J. Am. Chem. Soc., 101 (1979) 4554-4558.
22 F.S. Baker, A.M. Bradshaw, J. Pritchard and K.W. Sykes, Surface Sci., 12 (1968) 426-436.
23 M.J. Heal, E.C. Leisegang and R.G. Torrington, J. Catal., 51 (1978) 314-325.
24 D. Fraenkel and B.C. Gates, J. Am. Chem. Soc., 102 (1980) 2478-2480.
25 J.H. Lunsford, P.J. Hutla, M.J. Lin and K.A. Windhorst, Inorg. Chem., 17 (1978) 606-612.
26 H. Praliaud, G. Coudurier and Y. Ben Taarit, J. Chem. Soc., Faraday I, 74 (1978) 3000-3007.
27 H.H. Nijs, P.A. Jacobs and J.B. Uytterhoeven, J. Chem. Soc. Chem. Commun., (1979) 180-181.
28 D. Vanhove, P. Makambo and M. Blanchard, J. Chem. Soc. Chem. Commun., (1979) 605-606.

INFLUENCE OF CONTROLLED STRUCTURAL CHANGES ON THE CATALYTIC
PROPERTIES OF ZEOLITES

P. JÍRŮ

J. Heyrovský Institute of Physical Chemistry and Electrochemistry,
Czechoslovak Academy of Sciences, 121 38 Prague 2 (Czechoslovakia)

ABSTRACT

The paper is aimed at a detailed discussion of the results of
an experimental investigation of the effect of changes in surface
and/or crystallographic structure of zeolites of the types Y and
ZSM on their reactivity. The test interactions were:
i) the activation of hydrogen by the zeolites Fe(II)Y;
ii)the interaction of the dealuminated zeolites HY and that of the
zeolites ZSM with methanol.

INTRODUCTION

A series of new catalytic processes using zeolitic catalysts
with a higher content of Si has been realized recently in the
world. In order to explain from the theoretical point of view the
remarkable activity and selectivity of these zeolitic catalysts
the laboratories taking part in this research have published a
great amount of more generally valid data about their synthesis
and modification. The achieved results give suggestions for controlled changes both in the crystallographic and surface structure
of the zeolites (the former concerning e.g. the dimensions of the
cavities and their electrostatic field, the latter namely the type
and the amount of OH groups, the presence and the location of protons, the composition of the surface layers of the zeolite crystallites and the Si:Al ratio) as well as for interpretations of their
influence on the catalytic behaviour of cations in zeolites. In
the preparation of metal-loaded zeolite catalysts (e.g. by reduction) these cations may act as the precursors of the corresponding
metal particles. Such changes may then affect the structural modification in question and optimize   it from the point of view of a
preferential course of a certain reaction.

The aim of our 4 contributions on the occasion of this workshop is to present the results of investigations in this field,

both experimental and theoretical. The following investigations were performed: i) a quantum chemical study on the molecular level of the properties of model clusters of faujasites with a varying Si:Al ratio and various forms of Cr, Fe, Co and Ni; ii) an experimental study of the influence of the pretreatment of the zeolites on the valence state, the localization and possibly the coordination of the cations of Fe, Cr and Ni as well as on the structural stability and catalytic activity of the respective zeolite; iii) the characterization of the structure of HY zeolites with a high degree of dealumination and the comparison of their surface structure and reactivity with the zeolites H-ZSM.

A more detailed discussion of the results of i) and ii) is presented in three separate papers* in this workshop - see Beran et al., Wichterlová et al., Patzelová et al.. My own paper will be dedicated to a survey of some recent results concerning the point iii) and the catalytic activity of the zeolites Fe(II)HY and that of the high-silica zeolites. In order to obtain an information as full as possible a series of experimental techniques, such as IR spectroscopy, ESCA, ESR, X-ray diffraction analysis and TPD, have been used.

CATALYTIC ACTIVITY OF THE Fe(II)HY ZEOLITES IN THE ACTIVATION OF HYDROGEN

The Fe(III) ions, introduced by ionic exchange up to 23 %, occupied preferentially the cationic sites and the stability of zeolitic structure increased. Cationic Fe(III) exhibited considerable tendency to self-reduction: after thermal treatment in vacuo (670 K) the zeolites contained 75 % of Fe(II). Thermal treatment up to 1070 K in vacuo and/or in hydrogen was accompanied by the dehydroxylation of structural OH groups and by the migration of Fe(II) cations into the large cavities.

Iron-rich Fe(III)NH$_4$Y zeolite (63 % of ion exchange) contained an appreciable amount of the hydroxo-oxidic form of Fe(III) in addition to Fe(III) in cationic sites. This hydroxo-oxidic form was more resistant against self-reduction than the cationic one. In contrast to zeolites with low content of iron no movement of Fe(II) species was found at high pretreatment temperatures; the structural stability was lower and nonspecific Si-OH groups predominated over the structural ones.

In all zeolites a very small amount of metallic iron (less than 10 %) was found in the surface layers after hydrogen reduction at

820 K.

The changes of Fe content and Fe localization as well as of the zeolite structure stability and of the nature and number of OH groups affect the deuterium exchange which was investigated with regard to the zeolite pretreatment on Fe(II)HY zeolites. Both types of deuterium exchange ($D_2 + H_2$ and $D_2 + OH$) were found not to be influenced by added Fe up to 7 % of ion exchange. A higher amount of Fe(II) accelerated the $D_2 + OH$ exchange in the pretreatment temperature range of 670 - 820 K. This acceleration resulted probably from the presence of a sufficient number of Fe(II) in the vicinity of OH groups. In this way the dissociation of the OH bond was enhanced and the D_2 exchange with OH groups predominated. With further increase of the pretreatment temperature the number of OH groups substantially decreased and thus the probability of the interaction of activated hydrogen species leading to $D_2 + H_2$ exchange increased. The presence of Fe(II) in large cavities seems to catalyze this type of exchange more likely than the small amount of metallic iron on the zeolite surface. This can be seen from the comparison of zeolites with 23 and 63 % of Fe ion exchange: the former zeolite with high $D_2 + H_2$ exchange rate exhibited a considerable number of Fe(II) in large cavities and a smaller amount of metallic iron in the surface layer, the latter one had a lower amount of Fe(II) in large cavities, a higher concentration of surface Fe^0 and a very low $D_2 + H_2$ exchange rate. This behaviour of Fe-rich zeolite can be explained by its structural collapse.

COMPARISON OF STRUCTURE AND REACTIVITY OF DEALUMINATED ZEOLITES WITH THOSE OF THE H-ZSM ZEOLITES

In this part of my paper I would like to discuss the independent influence of changes in the surface or crystallographic structure of the zeolites (i.e. in the absence of cations of transition metals and therefore also without their influence) on the course of the catalytic reaction. This independent effect must be taken into account also in the case of metal-loaded zeolite catalysts, as these represent in many cases typical bifunctional catalysts.

The X-ray diffraction data together with the IR spectra of the skeletal vibrations have shown the good crystallinity of our zeolites of the type H-ZSM (Si:Al ratios ranging between 17 and 49)

and dealuminated type HY (Si:Al ratios ranging between 2.5 and
18). The dealuminated catalysts were prepared by the method published by Beyer (ref.1). The subsequent cationic exchange with NH_4Cl
was performed with all these zeolites in the usual way.

The dealumination of HY just mentioned is interpreted (ref.1)
by the reaction of the zeolite with $SiCl_4$ at elevated temperatures, where Al from the lattice is exchanged for Si under the formation of volatile $AlCl_3$ as well as of NaCl possibly of $Na(AlCl_4)$.
These products are formed in the cavities of the zeolite and - according to (ref.1) - probably mostly removed from them in the course of the washing of the product with water. The substitution mechanism suggested is in good agreement with our experimental results comparing the dependence of the unit cell dimensions and
that of the OH groups vibrations on the content of aluminium as
(Al/Si+Al) in the series of the dealuminated zeolites. A part of
aluminium released from the lattice, however, is probably maintained in the cavities as extra-lattice aluminium or even as cationic
Al^{3+}. The dealuminated zeolites are highly decationated, which is
probably the result of the hydrolysis of Cl complexes in the course of the preparation. The further exchange of Na^+ for NH_4^+ has a
minimum effect.

The content of strongly acidic structural groups in the dehydrated and deammonized Y zeolites decreases with the content of Al
whereas the amount of nonspecific SiOH hydroxyls (IR band at 3745
cm^{-1}) increases. In dealuminated HY zeolites hydroxyl groups corresponding to the IR band near 3600 cm^{-1} were observed: this band
indicates obviously the presence of extra-lattice aluminium, as it
has been found by many authors in the case of hydrothermally stabilized Y zeolites, where the delocalization of Al from the lattice into the cavities is typical.

The increase of the skeletal Si:Al ratio, the presence of extra-
-lattice aluminium and the decationization strongly influence the
acido-basic properties of dealuminated Y zeolites, their stability
and catalytic activity. In some cases also other factors may be
important: the high enrichment of the surface layers with Al and
the presence of traces of Cl in these layers, as determined by
X-ray photoelectron spectroscopy. In this respect, our results
complete some conclusions about similar zeolitic systems presented
in (ref.2).

The products of the interaction of methanol with H-ZSM and highly dealuminated HY zeolites, which are released successively with

increasing temperature, were investigated by the TPD technique
with mass-spectrometric analysis of the desorbates. The TPD measurements were performed in the temperature range 300 - 800 K.

The dependence of the relative ratios of saturated, olefinic
and aromatic desorbates (products of interaction) on the atomic
Si:Al ratios on dealuminated HY zeolites (with the same crystallographic but varying surface structure) will be discussed in detail.

On the whole, a comparable amount of dimethyl ether is formed on
all zeolite probes. The highest amount of olefins (namely of propylene) and of aromates appears on HY zeolites with the Si:Al ratio between 3 and 4; it is about twice as high as that observed on
HY zeolites with the Si:Al ratio of about 17. No paraffins (including methane) were found after the interaction of these zeolites
with methanol.

Further attention to differences in the reactivities of both
types of zeolites HZSM-5 and HY with the same ratio of Si:Al = 17
and different crystallographic structures is given. On the whole,
again a comparable amount of dimethyl ether is formed on both types
of zeolites. In contrast to the dealuminated HY zeolite, in the
interaction of methanol with the H-ZSM-5 zeolite among the paraffins mainly methane is formed and no olefins were observed. The
products of interaction contain comparable amounts of aromates.

REFERENCES
1 H.K. Beyer and I. Belenykya, Catalysis by Zeolites, Ed. B.
   Imelik et al., Elsevier, Amsterdam, 1980, 203 pp.
2 P.A. Jacobs, J. Weitkamp and H.K. Beyer, Faraday Discussion,
   1981, preprint 72/21.

BEHAVIOUR OF Fe SPECIES IN ZEOLITE STRUCTURE

B. WICHTERLOVÁ, L. KUBELKOVÁ, J. NOVÁKOVÁ and P. JÍRŮ

J. Heyrovský Institute of Physical Chemistry and Electrochemistry, Czechoslovak Academy of Sciences, 121 38 Prague 2 (Czechoslovakia)

ABSTRACT

The changes of the form and reducibility of Fe(III) ion complexes within the zeolite structure resulting from zeolite hydrothermal treatment and treatment with NaCl solution were studied by means of ESR spectra of Fe(III) ion, TPR by hydrogen, IR spectra of adsorbed CO and XPS spectra of zeolite surface layers. The parent FeNH$_4$Y and FeNaY zeolites contained Fe(III) in cationic sites mainly, where they were easily self-reduced by vacuo and reduced by hydrogen to Fe(II). Along with the reorganization of zeolite structure the hydrothermal treatment causes the formation of Td coordinated Fe(III) in Fe-O and Al-O clusters which are not self-reduced and are with difficulty reduced by hydrogen to Fe(II). NaCl solution treatment does not lead to a simple exchange of Fe(III) ions from cationic sites into the solution. However, their rearrangement into new, most likely hydrolytic complexes within the zeolite takes place. In contrary to Fe(III) in cationic sites, these complexes are not self-reduced by vacuo and are partially reduced by hydrogen to metallic Fe.

INTRODUCTION

The Fe(III), Fe(II) and Fe^O species localized within the zeolite structure are of a great and permanent interest in the field of zeolite chemistry and catalysis (refs.1-23). This is due to the fact that Fe(III) ions are present in various trace amounts in zeolites and that Fe ions, exchanged into the cationic sites as well as introduced in the form of oxidic clusters, appreciably modify zeolite catalytic activity (refs.1-3,13,15). Moreover, the reduced Fe species in zeolites are responsible for the activity in Fischer-Tropsch synthesis (refs.17,19). A lot of studies deal with the characterization of the redox properties of Fe ions, especially those in cationic sites (refs.4,8,11,12,20,22). It has been found that Fe(III) ions can be localized in the cationic positions of zeolites, being quite easily reduced to Fe(II) valence state. Further reduction of Fe ions in these sites to Fe^O does not take place without collapse of the zeolite structure.

Our previous papers (refs.11,12,20) concerned the characterization and reducibility of Fe(III) ions depending on their localization and form of their complexes. They are both influenced by the
conditions of Fe ions introduction into the zeolite and further
zeolite treatment. It has been shown that Fe(III) ions isomorphous-
ly substituting Al in zeolite framework are very resistant to reduction by hydrogen, on the other hand the Fe(III) in cationic sites are reduced very easily under vacuum or by hydrogen even
though to the Fe(II) valence state only. The behaviour of Fe(III)
in hydroxo-oxidic species is more complex and not fully explicit.
The vast amount of them is not self-reduced, but reduced by hydrogen to Fe(II). In contrary to pure Fe_2O_3, only a small part of
such complexes in zeolite surface layers is reduced to Fe^0.

This paper deals with the further changes of the character of
Fe(III) complexes within the zeolite structure as a result of the
following zeolite treatment: i) hydrothermal treatment, which is
well-known as a process of ammonium zeolites structure stabilization, and ii) treatment with sodium chloride solution. Both these
processes result in considerable changes of the form, localization
and thus reducibility of Fe(III) complexes.

EXPERIMENTAL

$FeNH_4Y$ (2.70 wt. % of Fe) was prepared from $(NH_4)_{95}Na_5Y$ zeolite
by Fe(III) ion exchange in such a way that Fe(III) ions were introduced predominantly into the cationic sites and the zeolite
crystallinity was retained (cf.refs.4,5,12). The hydrothermal
treatment of this zeolite, performed at 1030 K for 3 hours in flow
of oxygen saturated with 80 % of water vapour and subsequent hydration at 298 K, yielded stabilized well crystalline FeSY zeolite; details see in refs. 23,24. FeNaY (2.34 wt. %), prepared
from NaY zeolite by Fe(III) ion exchange, contained Fe(III) mainly
in cationic sites, too. Further zeolite modifications were performed by i) 0.1n NaCl solution applied twice (FeNaY-1) and
ten times (FeNaY-2) for 10 minutes at 298 K, and ii) 0.1n NaOH
solution for 1 hour at 298 K (FeNaY-3). The zeolites crystallinity
was tested by the X-ray pattern, IR spectra of skeletal vibrations
and sorption capacity of argon.

For the characterization of the form and reducibility of Fe(III)
complexes were used the ESR spectra of Fe(III) ions, the temperature programmed reduction by hydrogen (TPR), the IR spectra of adsorbed CO and XPS spectra of surface layers of zeolites. Details of

the methods, specifications of instruments and detail spectra interpretation were described elsewhere (refs.9,11,12).

Before the measurements the zeolites were heat treated at various conditions: in flow of oxygen at 670 K for 3-18 hours, in vacuo of 10^{-3} Pa at 670 K for 18 hours, sometimes followed by treatment with hydrogen at 0.5 kPa and at 670 or 820 K. The CO was adsorbed at the temperature of IR beam under a pressure of 6 kPa.

RESULTS AND DISCUSSION
Parent FeNH$_4$Y and FeNaY zeolites

Both the original FeNH$_4$Y and FeNaY zeolites contained Fe(III) ions predominantly in the cationic sites. This was demonstrated from the characteristic behaviour of such located Fe(III) ions (cf.ref.12), i.e. from the nearly reversible change of the ligand field symmetry from Oh to Td in the cycles hydration-dehydration and easy self-reduction by vacuo to Fe(II).

The above mentioned zeolites in hydrated form were characterized by the ESR signal intensity at g=2.0, ΔH=500 G of hydrated Fe(III) complexes with Oh symmetry of the ligand field, zeolites dehydrated in oxygen gave the signal at g=4.3, ΔH=50 G of dehydrated Td coordinated Fe(III) complexes (Fig. 1a; cf.refs.6,7,10-12). Such Fe(III) coordination in zeolites was determined by the Moessbauer and X-ray data, too (refs.8,18). The zeolites evacuated at elevated temperature did not exhibit ESR signal at g=4.3 of dehydrated Td Fe(III) complexes as a result of Fe(III) ions self-reduction. The consumption of hydrogen (in TPR spectra) for zeolites treated in vacuo was considerably lower than that for zeolites treated in oxygen (Fig. 2, Table). In addition, the zeolites treated in vacuo and/or hydrogen exhibited in the IR spectra of adsorbed CO band at 2197 cm^{-1} which indicates the Fe(II) ions in S_{II} cationic positions (Table and cf. ref.12 and refs. herein).

The Fe(III) ions in the framework positions as well as Fe(III) ions in hydroxo-oxidic species were present along with the Fe(III) in the cationic sites. The traces of the skeletal Fe(III) were evidenced by the presence of the low intensity ESR signal at g=4.3 in hydrated zeolites. The small amount of hydroxo-oxidic species, formed as a result of partial hydrolysis was identified by the broad signal with a maximum at g=2.3, ΔH=1300 G (cf.refs. 6,7,11,12,14).

Hydrothermal treatment of $FeNH_4Y$ zeolite

The hydrothermal treatment of $FeNH_4Y$ results in a substantial reorganization of zeolite structure including leaching of Al from skeleton into Al-O clusters in cavities, enrichment of surface layers with Al (refs.23,24) and in considerable changes of the form of Fe(III) complexes.

The character of TPR curves for FeSY evidences much lower amount of readily reducible Fe(III) in comparison with $FeNH_4Y$ zeolite (Fig.2, Table). The Fe(III) ions are reduced only partly to Fe(II) and the extent of their self-reduction is limited, too. The XPS spectra provide an information on the depletion of Fe in the surface layers of about three times. The binding energies show that Fe(III) are present in oxidized FeSY; its treatment in vacuo or hydrogen causes reduction of Fe(III) ions to Fe(II). The reduction to Fe^0 in surface layers was observed in amount less than 10 % (Fig. 1b, Table). No considerable difference was found in the reducibility of Fe(III) species in surface layers of FeSY and $FeNH_4Y$ zeolites (cf. spectra in ref.9). Even though it is possible to distinguish the Fe(III), Fe(II) and Fe^0 valence states their quantitative analysis is questionable. Moreover, the binding energies of Fe 2p levels are not sufficiently sensitive to the coordination of Fe ions (ref. 25) and therefore the character of Fe species in zeolite surface layers cannot be determined. On the other hand, the ESR spectra show the new forms of Fe(III) complexes in FeSY zeolite. In contrary to $FeNH_4Y$, hydrated FeSY exhibits a considerable signal intensity at g=4.3, ΔH=50 G with a shoulder at g=6-10 and a low intensity signal at g=2.08, ΔH=60 G (Fig. 1a, Table). The former signal represents Td coordinated Fe(III), the latter one some hydrated Fe(III) not specified in details (refs. 23,24). The changes of the signal intensity at g=4.3 caused by various FeSY treatments evidence that the new form of Fe(III) complexes is not self-reduced and also not reduced by hydrogen up to 670 K. Its reduction at 870 K and easy reoxidation at 670 K takes place. We observed a similar redox behaviour of Fe(III) in skeletal positions of alumina.

It can be assumed that the new form of Fe(III) complexes responsible for the lower reducibility of FeSY is Td coordinated Fe(III) built into Al-O as well as Fe-O clusters strongly bound to zeolite skeleton. This suggestion is based also on the fact that it was possible to dissolve only some of these Fe(III) from zeolite along with Al-O clusters (see ref.23).

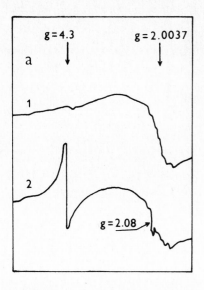
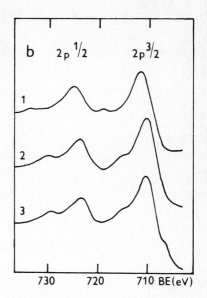

Fig. 1. ESR and XPS spectra of Fe-zeolites
(a) X-band ESR spectra at 298 K of 1) FeNH$_4$Y, 2) FeSY. (b) XPS
spectra of Fe 2p levels of FeSY treated in oxygen at 670 K and
measured 1) as rec., 2) evac. at 670 K, 3) in hydrogen at 820 K

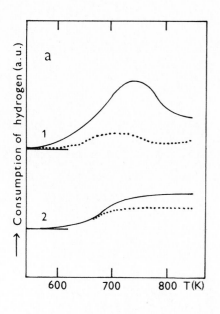
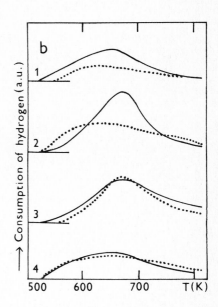

Fig. 2. TPR curves of Fe-zeolites
Zeolites pretreated at 670 K in oxygen ——, in vacuo ....
(a) 1) FeNH$_4$Y, 2) FeSY. (b) 1) FeNaY, 2) FeNaY-1, 3) FeNaY-2,
4) FeNaY-3

Some Fe(III) remain present in cationic sites of FeSY. It is evident form the difference between i) the signal intensities at g=4.3 for FeSY hydrated, treated in vacuo and oxygen, and ii) consumptions of hydrogen for FeSY treated in oxygen and vacuo (Table). However, no IR bands of adsorbed CO on Fe(II) in the cationic sites were found for FeSY either treated in vacuo or in hydrogen. Thus we assume that the cationic Fe(III), observed by ESR and TPR, are located either in small cavities or they are blocked by Al-O and/or Fe-O clusters; therefore Fe(II) are not accessible for CO molecules. Finally, the amount of hydroxo-oxidic species, checked by the broad ESR signal at g_{max}=2.3, does not increase due to hydrothermal zeolite treatment.

TABLE
Characterization of zeolites

Zeolite	treated at 670K in	consumption of hydrogen[a] (μmol/g)	A(CO)[b] (cm^2/g)	g=4.3[c]	Fe $2p^{3/2}$ (eV)	sat. (eV)	Δ[d] (eV)
FeNH$_4$Y	oxygen	257	-	90	711.7	8.3	13.7
	vacuo	95	4.8	3	710.5[e]	5.0	13.1
	hydrogen		4.8	3	710.5[e]	5.0	13.3
FeSY	oxygen	129	-	150	711.7	8.0	13.5
	vacuo	105	0.0	75	710.7[e]	4.5	13.5
	hydrogen		0.0	70	710.6[e]	4.5	13.3

[a]Theoretical consumption of hydrogen for the reduction of Fe(III) to Fe(II) is 257 μmol/g of dry bases.
[b]Normalized height of 2197 cm^{-1} band.
[c]Peak to peak height.
[d]BE are related to Si 2p level of 102.8 eV.
[e]Zeolites were treated in vacuo at 770 K and in hydrogen at 820 K.

Natrium chloride solution treatment of FeNaY zeolite

The chemical analysis of NaCl treated zeolites brought an evidence that all Fe(III) ions remain present within the zeolite. However, the surface layers are of about twice enriched by Fe(III), as follows from the XPS spectra of Fe 2p levels, even though the major part of them remains in the bulk. As the form of Fe(III) is concerned its character was changed considerably with respect to reduction in comparison with parent FeNaY. The TPR curves for FeNaY-1 and -2 zeolites evidence progressive decrease of Fe(III) ions selfreducibility (Fig. 2b). Moreover, the hydrogen consumption exceeds the theoretical amount of hydrogen necessary for the re-

duction of Fe(III) to Fe(II) ions and corresponds to reduction of
a part of Fe(III) to Fe^o. The extent of this reduction is 20 and
30 % of Fe(III) for FeNaY-1 and -2, respectively. While the pure
water treatment has no effect on the reducibility as well as other
characteristics of cationic Fe(III), zeolite treatment with NaOH
solution changes also the form of Fe(III) complexes; the Fe(III)
ions of FeNaY-3 loose the ability to be self-reduced, but the consumption of hydrogen corresponds to the theoretical reduction of
Fe(III) to Fe(II).

It is obvious that both NaCl and NaOH solutions attack Fe(III)
ions in cationic sites in contrast to water, due to their ability
to exchange cations. The following step seems to be an immediate
hydrolysis, precipitation of Fe(III) ions and the formation of
some hydrolytic species inside zeolite cavities. As the character
of these species is different for zeolites treated with NaCl or
NaOH solutions, the new forms of Fe(III) must be strongly influenced by the medium in cavities.

Recently Scherzer and Fort (ref.16) published the method for
the preparation of metal particles within the zeolite. It is based
on the precipitation reaction of Fe ions in cationic sites with a
water soluble and Fe containing anionic coordination compound and
subsequent reduction of this precipitate. The method described
here offers other and simple way of the preparation of Fe metal
within zeolite with the preservation of zeolite structure.

CONCLUSION

The Fe(III) ion can be localized in many forms of complexes and
positions of zeolite structure. Fe(III) can substitute Al in zeolite framework, it can be localized in cationic sites as well as
in Al-O and Fe-O clusters and form various types of hydroxo-oxidic
species. The form of Fe(III) complex determines strongly the
Fe(III) ion reducibility and therefore the final valence state of
Fe species within the zeolites. The character of Fe(III) complexes
is influenced not only by the manner of Fe(III) ions introduction
into zeolite but significantly by the further zeolite treatment.
During the hydrothermal treatment the Fe(III) ions are built in
Al-O and Fe-O clusters which are formed as a result of zeolite skeleton reorganization. In this form Fe(III) are with difficulty reduced even by hydrogen. Treatment of zeolite with natrium chloride
solution changes the form of Fe(III) in cationic sites to special
hydrolytic species which are reduced to metallic Fe without the

collapse of zeolite structure.

REFERENCES

1   P.A. Jacobs, Carboniogenic Activity of Zeolites, Elsevier, Amsterdam, 1977.
2   N.N. Bobrov, G.K. Boreskov, K.G. Ione, A.A. Terletskikh and N.A. Schestakova, Kin. Kat. 15 (1974) 413.
3   C.F. Heylen, P.A. Jacobs and J.B. Uytterhoeven, J. Catal. 43 (1976) 99.
4   J.A. Morice and L.V.C. Rees, Trans. Faraday Soc. 64 (1968) 1388.
5   B. Wichterlová, L. Kubelková, P. Jirů and D. Kolihová, Coll. Czech. Chem. Commun. 45 (1980) 2143.
6   B.D. McNicol and G.T. Pott, J. Catal. 25 (1972) 223.
7   E.G. Derouane, M. Mestdagh and L. Vielvoye, J. Catal. 33 (1974) 169.
8   R.L. Garten, W.N. Delgass and M.J. Boudart, J. Catal. 18 (1970) 90.
9   P. Mikušik, T. Juška, J. Nováková, L. Kubelková and B. Wichterlová, J. Chem. Soc., Faraday Trans. 1, 77 (1981) 1179.
10 B. Wichterlová and P. Jirů, React. Kin. Catal. Letters 13 (1980) 197.
11 B. Wichterlová, Zeolites 1 (1980) 181.
12 J. Nováková, L. Kubelková, B. Wichterlová, T. Juška and Z. Dolejšek, Zeolites, 2 (1982) 17.
13 L. Kubelková and J. Nováková, J. Mol. Catal., in press.
14 W. Gunsser, F. Schmidt, G.N. Belozerskij and M. Kasakov, Proc. XXth Congress Ampere, Tallin (1978), p. 251.
15 T. Imai, and H.W. Hagbood, J. Phys. Chem. 77 (1973) 925.
16 J. Scherzer and D. Fort, J. Catal. 71 (1981) 111.
17 V.U.S. Rao, J.R. Gormley, Hydrocarbon Processing, Nov. (1980) 139.
18 J.R. Pearce, W.J. Mortier, J.B. Uytterhoeven and J.H. Lundsford, J. Chem. Soc., Faraday Trans. 1, 77 (1981) 937.
19 D. Ballivet-Tkatchenko and I. Tkatchenko, J. Mol. Catal. 13 (1981) 1.
20 S. Beran, P. Jirů and B. Wichterlova, Zeolites, in press.
21 Y.Y. Huang and J.R. Anderson, J. Catal. 40 (1975) 143.
22 F. Schmidt, W. Gunsser and J. Adolph, ACS Symp. Ser. No. 40 (1977) 291.
23 B. Wichterlová, L. Kubelková, J. Nováková and P. Mikušik, submitted for publication.
24 B. Wichterlová, J. Nováková, L. Kubelková and P. Jirů, Proc. 5th Int. Conf. Zeolites, Ed. L.V.C. Rees, Heyden, London, (1980) p. 373.
25 C.R. Brundle, T.J. Chuang and K. Wandelt, Surface Sci. 68 (1977) 459.

TO THE DIFFERENCES IN PROPERTIES OF Ni METAL PARTICLES AND Cr CATIONS IN STABILIZED AND NONSTABILIZED ZEOLITES

V. PATZELOVÁ, Z. TVARŮŽKOVÁ, K. MACH and A. ZUKAL

J. Heyrovský Institute of Physical Chemistry and Electrochemistry, Czechoslovak Academy of Sciences, Máchova 7, 121 38 Prague 2, Czechoslovakia

ABSTRACT

Series of stabilized Y zeolites were prepared differing in the residual sodium ion content and in the temperature of their hydrothermal treatment. The properties of the Ni^{2+} and Cr^{3+} ions exchanged into the above mentioned structures were compared with the properties of Ni^{2+} and Cr^{3+} ions located in nonstabilized Y zeolites of the same chemical composition.

In the case of nickel the stabilized structure causes an increase in the reducibility of Ni^{2+} ions. As follows from the comparison of the results with those published by other authors (ref.1), small monodispersed metal particles are created during the reduction of the Ni^{2+} ions located in the stabilized structures of Y zeolites. The resulting samples exhibit high catalytic activity in the CO reduction. Increased catalytic activity was also observed in the case of Cr ions located in the stabilized structure.

The electron paramagnetic and infra-red spectroscopic measurements have shown that the Cr^{3+} and Cr^{5+} ions are active in the ethylene oligomerization. A high degree of crystallinity of the stabilized catalysts is maintained even after repeated regeneration.

INTRODUCTION

Stabilized forms of Y zeolites, their synthesis and properties have been first described by McDaniel and Maher (ref.2). Since that time several methods of their preparation have been reported (ref. 3,4,5,6,7).

It has been shown in our previous paper (ref.8) that the stabilized structure of Y zeolite can be obtained even in samples containing up to 65 % of Na^+ cations. Such forms have the faculty of combining the properties of stabilized structure with the high ion exchange capacity. During the process of stabilization a relocalization of the metal cations takes place. The main part of Na^+ cations migrates into the supercages, as was stated by the values of the adsorption heats of CO and C_2H_4 and by IR measurements (ref.9). Simultaneously up to 50 % of the Al atoms are removed from the framework. These effects influence the location of Ni^{2+} and Cr^{3+} ions exchanged subsequently (ref.10).

The present study was undertaken to confirm the positive influence of the stabilized structure of Y zeolites on the catalytic properties of Ni and Cr cations.

EXPERIMENTAL

The starting substance was NaY zeolite supplied by the Institute of Petroleum and Hydrocarbon Gases, ČSSR. The SiO_2/Al_2O_3 molar ratio of this material was 4.6. The standard ion-exchange procedure was used for the preparation of the $NaNH_4Y$ forms. The process of stabilization was carried out at 823 and 1043 K, respectively. So far as the stabilization treatment is concerned, we have given full details of the experimental procedure earlier (ref.11). The ion-exchange of the residual Na^+ ions in all samples was made in 0.05 N solutions of Ni^{2+} or Cr^{3+} chlorides. The samples with Ni^{2+} ions were afterwards dehydrated at 625 K in dried helium for 3 hrs and then reduced in the stream of dried electrolytic pure hydrogen at a specified temperature between 653 and 717 K. The chemical characterization of studied samples is given in Table 1. The content of Ni^{2+} and Cr^{3+} ions was obtained by the analysis of the solid phase. The degree of reduction was determined from the amount of nickel ions eluted during the back ion-exchange of the partially reduced samples with a 0.1 N solution of $CaCl_2$.

TABLE 1

Characterization of the catalyst samples

Sample No.	Ni (wt %)	Cr (wt %)	Degree of decationization (%)	Temperature of stabilization (K)
1	4.10	-	47	823
2	3.64	-	47	1 043
3	3.96	-	47	-
4	6.10	-	47	823
5	-	1.61	31	823
6	-	2.75	68	-

The ferromagnetic resonance (FMR) and the electron paramagnetic resonance (EPR) measurements were performed with the ESR-220 spectrometer (Zentrum für wissenschaftlichen Gerätebau, Akademie der Wissenschaften der DDR) at the range of temperatures 89-296 K. Before the FMR measurements the nickel containing samples were reduced in suitable glas capillaries. After the reduction the hydrogen was removed by helium and the capillaries were sealed. The evaluation of the FMR spectra was performed analogically to the methods described by D. Olivier "et al". (ref.12).

The infrared (IR) spectra were recorded with a Beckman IR-7 spectrometer. The preparation of self-supporting wafers for IR measurements was published earlier (ref.13). All details of their pretreatment are given in the text to the Figure 3.

The catalytic activity of nickel samples was tested in a simple pulse reactor combined with a gas chromatograph.

The catalytic properties of chromium ions were tested in the oligomerization of ethylene. The activity of each investigated catalyst was considered to be proportional to the weight increase of the tested sample (measured on quartz balance); the weight increment was caused by the formation of high-molecular polyethylene. The details of the procedure were published (ref.13).

RESULTS

In the first part of the experiments the influence of the stabilized structure on the reducibility of Ni^{2+} ions was followed.

TABLE 2
The reducibility of Ni^{2+} ions in dependence on the temperature

Temperature of reduction (K)	Sample No.	Degree of reduction[a] (wt %)	Temperature of reduction (K)	Sample No.	Degree of reduction[a] (wt %)
658	1	46.8	697	1	47.1
	2	55.8		2	59.6
	3	11.5		3	13.2
677	1	44.7	717	1	49.0
	2	56.0		2	67.7
	3	12.2		3	15.5

[a]Degree of the nickel reduction determined from the results of the back-exchange with $CaCl_2$ solution. Time of reduction 10 hrs.

We may see that the concentration of Ni particles at each temperature is higher in stabilized structures. The deeper are the structural changes caused by the stabilization (see ref.8) the higher is the degree of reduction.

The results of FMR measurements performed in the range of temperatures 143-296 K enabled to characterize the shape and dispersity of created metal particles. The g factors of stabilized catalysts

were calculated from the position of isosbestic point (ref.1).
The existence of this point is a criterion of the monodispersion
of small metal particles. As there was no isosbestic point in the
case of nonstabilized samples, their g factors were calculated
from the positions of H_o. The results are given in Table 3. The
monodispersion of small Ni particles in the stabilized samples was
also confirmed by the linear decrease of the half-widths ΔH_{pp} of
the FMR lorenzian lines with increasing temperature.

TABLE 3
The g values of Ni^o particles and the activation energies of Ni
catalysts in the carbon monoxide reduction

Sample No.	g factor	Activation energy $(kJ.mole^{-1})$
1	2.22	63.9 ± 1.7
2	2.17	64.7 ± 2.1
3	2.23	68.8 ± 1.4
4	2.21	64.1 ± 1.4

The catalytic activities of the studied materials were tested
in the reaction of carbon monoxide reduction. The stabilized catalysts 1, 2 and 4 were completely reduced after 20 hrs treatment in
H_2 at 673 K. In sample 3 reduced for 30 hrs at 673 K 50 % of Ni^{2+}
ions remained unreduced.

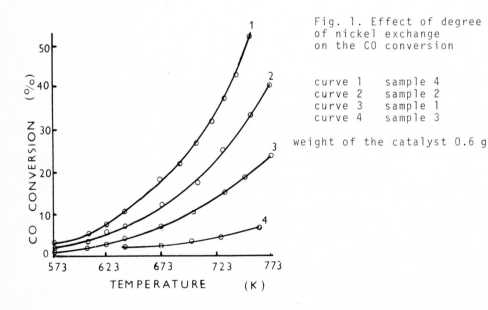

Fig. 1. Effect of degree
of nickel exchange
on the CO conversion

curve 1   sample 4
curve 2   sample 2
curve 3   sample 1
curve 4   sample 3

weight of the catalyst 0.6 g

The nearly equal values of activation energies obtained for the stabilized samples differing in the content of nickel indicate that the increased metal concentration influences only the quantity but not the quality of active centra.

The influence of the stabilized structure on the catalytic activity of Cr ions was tested in the ethylene oligomerization.

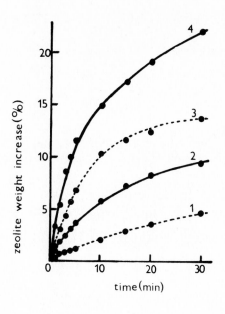

Fig. 2. Kinetic curves of ethylene oligomerization on zeolites at 5.3 kPa and at 350 K

Sample 5 (———), sample 6 (-----).
Pretreatment conditions:
1) and 2) evac. 620 K 12 hours;
3) and 4) evac. 620 K 12 hours
       then oxidation 770 K
       3 hours and evac.
       620 K 1 hour

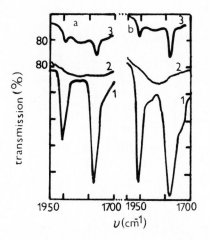

Fig. 3. IR spectra of zeolites after adsorption of NO at 2.7 kPa and at 298 K

Sample 5 - part a, sample 6 - part b.
Pretreatment conditions:
1) evac. 620 K 1 hour;
2) evac. 620 K 12 hours then oxidation 770 K 3 hours and evac. 620 K 1 hour;
3) evac. 620 K 12 hours then oxidation 770 K 3 hours and evac. 770 K 1 hour

As follows from the comparison of curve 1 with curve 2, the stabilized catalyst is more active than the nonstabilized one. After the oxidation under the given conditions the activity of both samples is higher but the proportionality between them is preserved.

In order to obtain more information about the catalytic active valence states of Cr cations, IR spectra of adsorbed nitrogen oxide were recorded before and after the oxidation.

The bands at 1900 and 1780 cm^{-1} (see spectra 1a and 1b) can be ascribed according to the literature (ref.14,15) to the valence vibrations of the nitric oxide bound in the complex $Cr^{3+}(NO)_2$. After the oxidation at elevated temperature the bands disappear (see spectra 2a,2b) because of the increase of the valence state of Cr ions, probably to Cr^{5+}. As follows from the spectra 3a and 3b the evacuation of the oxidized samples at 770 K causes a partial self--reduction of Cr^{5+} ions and creation of Cr^{3+} ions.

The presence of Cr^{3+} ions in stabilized and nonstabilized structures (sample 5, 6) was stated by the EPR signal of the $Cr^{3+}(NO)_2$ complex at $g_\perp$ = 1.986 and g_{II} = 1.907. Both oxidized samples gave an EPR signal with ΔH_{pp} = 50 G and g = 1.978 characteristical for Cr^{5+} ions. The intensity of the Cr^{5+} signal decreases rapidly after the evacuation at 770 K, the decrease being caused by the self-reduction of Cr^{5+} ions.

It is worth noting that there is no qualitative difference between the behaviour of Cr ions in stabilized and in nonstabilized structures so far as the valence state changes are concerned. The proportional increase of the catalytic acitivities of both catalysts observed after the oxidation indicates that Cr^{5+} ions are more active in the ethylene oligomerization than the Cr^{3+} ions. The sensitivity of the higher valence state to self-reduction is about the same for both samples.

DISCUSSION

As has been stated, the reducibility of nickel cations in zeolites can be increased by the presence of previously incorporated rare earth cations or by pretreatment with carbon monoxide molecules at an elevated temperature (ref.12). The purpose of these operations is to increase the number of Ni^{2+} ions in open S_{II} and $S_{II'}$ positions. As has been shown (ref.16) a linear relationship exists between the degree of nickel reduction and the nickel concentration in S_{II} and $S_{II'}$ positions. In the dehydrated NiY zeolites the nickel ions in S_I sites are coordinated octahedrally with six structural oxygens; in $S_{II'}$ and S_{II} positions they are coordinated trigonal-pyramidally. It is conceivable that the Ni^{2+} ions in S_I sites were reduced with great difficulty because of their most stable octahedral coordination, as was verified by Briend-Faure "et al." (ref.17). It can be concluded that the reducibility of Ni^{2+} cations is predominantly influenced by their location in the zeolitic matrix.

As has been mentioned earlier, during the preparation of our stabilized Y zeolites the residual Na^+ cations migrate into the accessible positions and a part of the Al species created is situated in the small cavities. From the adsorption measurements follows that the main part of Ni and Cr ions exchanged into the stabilized structures is situated in positions which are accessible to the carbon monoxide and ethylene molecules (ref.9,10). In the nonstabilized structures of the same chemical composition the Ni and Cr cations occupy predominantly the unaccessible positions.

On the basis of the presented results the increased reducibility of Ni^{2+} ions and higher activity of nickel and chromium stabilized catalysts is explained by the increased number of the mentioned cations or active centres in positions accessible to the molecules of the reactant. The increased catalytic activity of both Cr catalysts after their oxidation is ascribed to the higher activity of Cr^{5+} ions in the ethylene oligomerization and not to an increased number of active centra.

REFERENCES

1   D. Olivier, M. Richard, L. Bonneviot and M. Che, in J. Bourdon (Ed.), Growth and Properties of Metal Clusters, Elsevier, Amsterdam, 1980, p. 193.
2   C.V. Mc Daniel and P.K. Maher, in Proc. 1st Int. Conf. Molecular Sieves, London, April 4-6, 1976, Soc. of Chem. Ind., London, 1968, p. 186.
3   G.T. Kerr, J. Phys. Chem., 71 (1967) 4155.
4   G.T. Kerr, J. Phys. Chem., 72 (1968) 2594.
5   G.T. Kerr, J. Catalysis, 15 (1969) 200.
6   J. Scherzer and J.L. Bass, J. Catalysis, 46 (1977) 100.
7   K.H. Beyer and J. Belenykaja, in B. Imelik, C. Naccache, Y. Ben Taarit, J.C. Vedrine, G. Coudurier and H. Praliand (Eds.), Proc. Int. Symp. Catalysis by Zeolites, Lyon, September 9-11, 1980, Elsevier, Amsterdam, 1980, p. 203.
8   V. Patzelová, V. Bosáček and Z. Tvarůžková, Acta Phys. et Chem. Szeged, XXIV (1978) 257.
9   V. Patzelová, V. Bosáček and Z. Tvarůžková, in Proc. of the Workshop Adsorption of Hydrocarbons in Zeolites, Berlin, November 19-22, 1979, AdW DDR, Berlin, 1979, p. 148.
10 V. Patzelová and Z. Tvarůžková, React. Kinet. Catal. Lett.,

16 (1981) 65.

11 Z. Tvarůžková, V. Patzelová and V. Bosáček, React. Kinet.
    Catal. Lett., 6 (1977) 433.

12 M. Che, M. Richard and D. Olivier, J. Chem. Soc., Faraday
    Trans., I 76 (1980) 1526.

13 Z. Tvarůžková and V. Bosáček, Coll. Czech. Chem. Commun., 45
    (1980) 2499.

14 J.R. Pearce, D.E. Sherwood, M.B. Hall and J.H. Lunsford,
    J. Phys. Chem., 84 (1980) 3215.

15 A. Zecchina, E. Garrone, C. Morterra and S. Coluccia, J. Phys.
    Chem., 79 (1975) 978.

16 M. Suzuki, K. Tsutsumi and H. Takahashi, Zeolites, 2 (1982)
    51.

17 M. Briend-Faure, J. Jeanjean, M. Kermarec and D. Delafosse,
    J. Chem. Soc., Faraday Trans., I 74 (1978) 1538.

INCORPORATION OF VOLATILE METAL COMPOUNDS INTO ZEOLITIC FRAME-
WORKS

P. FEJES, I. KIRICSI and I. HANNUS
Appl. Chemistry Dept., József Attila Univ., Szeged (Hungary)

ABSTRACT

Preparation methods were developed for introducing different
O- or halide-containing metal moieties into zeolitic frameworks.
The transformation of Brönsted acidity to Lewis acidity by
the reaction

$$\left[AlO_2^-\right]MCl_2^+ \xrightarrow{+H_2O} \left[AlO_2^-\right]MO^+ + 2HCl$$

deserves special attention from the point of view of catalytic
cracking, as no oxygen vacancies are produced simultaneously (see
Jacobs and Beyer [1]).
The halide ion-containing metal moieties exhibit noteworthy
activity in cyclopropane isomerization. The specimens treated
with $SnCl_4$ are hyperactive in this reaction, revealing that
allyl cations may be the active centres of the transformation.
At elevated temperatures dealumination takes place; this does
not necessarily lead to complete structural collapse if special
care is taken.

INTRODUCTION

There are a few papers in the literature describing the surface treatment of Aerosil (SiO_2) samples with volatile metal
halides [2-6] and alkyls [7-9] at relatively low temperatures
($\sim$ 600 K). It is presumed that the interactions taking place may
be characterized by the following stoichiometric equations

$$S-OH + MX_n \longrightarrow S-OMX_{n-1} + HX \qquad (1)$$

and

$$\begin{array}{l} S-OH \\ + MX_n \longrightarrow {}^{S-O}_{S-O}{>}MX_{n-2} + 2HX \\ S-OH \end{array} \qquad (2)$$

where S is a surface atom, M a metal atom, and X either a halide
atom or an alkyl group. Aerosil samples modified with $AlCl_3$, BCl_3
and $AlMe_3$ were tested in the cracking reaction of cumene (Hambleton et al. [6]) with the result that the activities were found to
surpass that of the untreated Aerosil, but to lag behind those of
the commercial aluminium silicates exhibiting Brönsted acidity.

A previous publication [10] has dealt with the dealumination of zeolitic samples using metal halides at elevated temperatures (above 700 K). It has been conjectured that prior to dealumination a halogen-containing moiety is incorporated into the zeolite according to the equation

$$\left[AlO_2^-\right]H^+ + MCl_3 \longrightarrow \left[AlO_2^-\right]MCl_2^+ + HCl \tag{3}$$

and in the form of a doubly charged species:

$$\left[AlO_2^-\right]MCl_2^+ + H^+ \longrightarrow \left[AlO_2^-\right]MCl^{2+} + HCl \tag{4}$$

where the symbol $\left[\ldots\right]$ denotes framework constituents. At higher temperatures, characteristic of the ions in question, a very interesting process takes place: a framework O^{2-} ion from among the nearest neighbours of the aluminium leaves the lattice to join M; thereafter the halide ions undergo rearrangement leading to the production of oxy-halide clusters and a framework vacancy ("empty nest"):

$$\left[AlO_2^-\right]MCl^{2+} \longrightarrow AlOCl + MOCl + \left[\ldots\right] \tag{5}$$

In this paper the catalytic properties of zeolitic samples were tested after their surface treatment with volatile metal halides, alkyls and alcoholates within the temperature range of stability of the incorporated moieties. The degree of dealumination too was determined in the case of samples exposed to higher pretreatment temperatures than the respective limit of stability.

EXPERIMENTAL

The zeolitic samples used in these experiments were $NH_4(Na)Y$ (Linde) and $NH_4(Na)$-mordenite (Norton) prepared from the Na forms by ion exchange and heat-treated prior to the modifications.

The surface treatment of the zeolite samples was carried out using solutions of the reactants in suitable solvents (method A) or gaseous substances introduced onto the zeolitic surface in a vacuum apparatus at low (mostly ambient) temperatures (method B).

Another way of contacting the zeolite with the reactant consisted in purging the sample at higher temperatures with an indifferent carrier gas (mostly N_2) containing the agent in low concentration (method C).

The amount of aluminium released in the process of dealumination was determined by means of standard procedures.

Possible structural changes in the modified specimens were checked via i.r. spectroscopy and X-ray diffraction.

Whereas the catalytic activities of the Y zeolite samples were tested in the skeletal isomerization reaction of cyclo-

propane, the cracking of propane was chosen as the test reaction
for the mordenite specimens. The kinetic investigations were carried out in a recirculatory flow reactor with g.c. product analysis.

RESULTS

a) <u>Zeolitic samples modified at near ambient temperatures</u>

H-mordenite treated with $AlCl_3$ according to method A (5 g H-
mordenite refluxed with 1.33 g $AlCl_3$ in 50 cm^3 ether in N_2 atmosphere until no more HCl evolution could be observed) resulted in
a dry sample after careful evaporation of the solvent and drying
at 393 K. The product was very carefully wetted with water vapour
in a desiccator for 24 hr, thereafter washed, and subsequently
dried at 393 K. Figure 1. shows the i.r. spectra of surface-trea-
ted mordenite specimens in the region of lattice vibrations (a)
prior to and (b) after wetting with water. The spectra are exactly the same, without any sign of (partial) lattice collapse.
The same conclusion can be drawn from the respective simplified
X-ray diffractograms in Fig. 2.

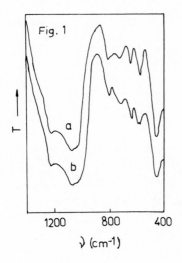
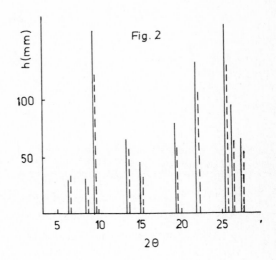

Fig. 1. Ir spectra of the treated samples: (a) prior to and (b)
after wetting with water.
Fig. 2. Simplified X-ray diagrams of the samples (solid line:
original sample, dashed line: treated mordenite after wetting
with water).

Another sample of H-mordenite was treated with Al(Et)$_3$ in a vacuum apparatus at 473 K. The release of ethane (as monitored via the pressure change)   is a good measure of the extent of reaction. After evacuation the fixed Al-alkyl was burned off at 773 K in O$_2$ atmosphere to yield an AlO$^+$ species in the exchange position. The rate of propane cracking (as revealed by the kinetic curves of the main reaction products in Fig. 3) was higher for the treated sample (b) than for the original specimen (a). A similar improvement in catalytic activity is observable in the case of dealuminated mordenites treated with Al(Et)$_3$. These results whow the important role played by Al$_2$O$_3$ in cracking, even in a very dispersed form.

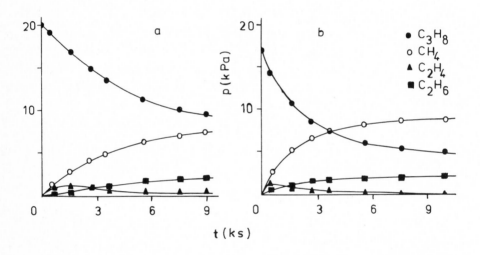

Fig. 3. Kinetic curves observed in the cracking of propane over the original NaH-mordenite (a) and the NaH-mordenite treated with Al(Et)$_3$ (b). (Reaction temperature 773 K; mass of catalyst 0.35 g)

Similar treatment of H(Na)Y with AlCl$_3$ in ether or CCl$_4$ leads to similar results as in the case of H-mordenite (the retention of crystallinity was as high as 92% with the CCl$_4$ solvent). It is very interesting to note that a nearly quantitative change of the surface Brönsted acidity into Lewis acidity caused a substantial <u>decrease</u> in the rate of skeletal isomerization of cyclopropane (see Fig. 4).

The explanation of the reversal of catalytic activity in the case of propane cracking and cyclopropane isomerization might be very complex in nature: topological effects (the transformation of Brönsted sites into sites of Lewis acidity is not quantitative when $Al(Et)_3$ is used) and slight differences in the mechanisms of cracking and skeletal isomerization, the details of which are not yet fully understood, might be operative.

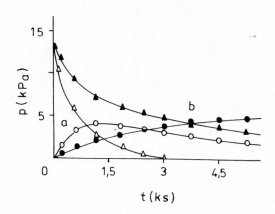

Fig. 4. Kinetic curves of cyclopropane isomerization observed at the untreated (a) and the treated (b) HY zeolite. (Reaction temperature 373 K, mass of catalyst 0,3 g; $\triangle$,$\blacktriangle$ cyclopropane, $\circ$,$\bullet$ propylene).

b) <u>Surface treatment at elevated temperatures</u>

Method C applied with volatile metal halides above 650 K leads to dealumination with only a minor loss of crystallinity. The extent of dealumination depends on several factors [type of zeolite and reactant (thermodynamics, pore size and dimensions of reacting molecules), and temperature]; a few illustrative examples are summarized in Table 1. It is worth noting that Beyer attained almost complete dealumination in the case of HY, using $SiCl_4$ [11].

By reducing the temperature of surface modification below a critical limit (dependent on both zeolite and reactant), it is possible to reduce and even fully avoid dealumination. For samples treated with metal halides below 673 K (method C) an improvement in the catalytic activity in cyclopropane isomerization can be observed as compared with the untreated HY. The sample treated with $SnCl_4$ is hyperactive (full conversion of cyclopropane into propylene <u>in a few seconds</u>, with a substantial rise in reaction temperature). This exceptional activity can be rationalized in terms of production of allyl cations (regarded as active centres of cyclopropane isomerization), which undergo oxidation in an autocatalytic fashion:

$$\left[\text{AlO}_2^-\right]\cdot\text{Sn}^{4+}\cdot(\text{OH})_2^{2-}\cdot\text{Cl}^- \xrightarrow{\;+\;\triangle\;} \left[\text{AlO}_2^-\right]\cdot\text{Sn}^{4+}\cdot\text{O}^{2-}\cdot(\text{OH})^-\cdot\text{Cl}^- \cdot \underset{\text{CH}_2^-\text{CH}_2}{\overset{\text{CH}_2}{/\text{H}^+\backslash}} \longrightarrow$$

non-classical carbonium ion

$$\xrightarrow[\text{oxidation}]{\;-\text{H}_2\text{O}\;} \left[\text{AlO}_2^-\right]\cdot\text{Sn}^{2+}\cdot(\text{OH})^-\cdot\text{Cl}^- \cdot \underset{\text{CH}\overset{+}{=}\text{CH}_2}{\overset{\text{CH}_2}{/\!\!\!/}} \xrightarrow[\text{isomerization}]{\;+\;\triangle\;}$$

allyl cation as active centre

$$\longrightarrow \left[\text{AlO}_2^-\right]\cdot\text{Sn}^{2+}\cdot(\text{OH})^-\cdot\text{Cl}^- \cdot \underset{\text{CH}\overset{+}{=}\text{CH}_2}{\overset{\text{CH}_2}{/\!\!\!/}} + \text{CH}_2 = \text{CH-CH}_3 \tag{6}$$

The water produced in the oxidation leads to the production of new Brönsted centres

$$\left[\text{AlO}_2^-\right]\cdot\text{Sn}^{4+}\cdot\text{Cl}_3^{3-} + \text{H}_2\text{O} \xrightarrow{\;-\text{HCl}\;} \left[\text{AlO}_2^-\right]\cdot\text{Sn}^{4+}\cdot(\text{OH})^-\cdot\text{Cl}_2^{2-} \tag{7}$$

the acidity of which is increased substantially in the presence of the electrophilic halide ions. Hydrolysis of Cl^- ions by water vapour reduces the activity.

TABLE 1

Amounts of framework constituents removed by volatile metal halides

Reagent	Temperature (K)	Al^{3+} (mmol g^{-1})	Fe^{3+} (μmol g^{-1})
TiCl_4	873	0.8	–
HgCl_2	773	0.95	13.75
FeCl_3	773	0.1	–

REFERENCES

1   P.A. Jacobs and H.K. Beyer, J. Phys. Chem., 83 (1979) 1174.
2   J.B. Peri, J. Phys. Chem., 70 (1966) 2937.
3   J.B. Peri and A.L. Hensley, J. Phys. Chem., 72 (1968) 2926.
4   G.C. Armistead, A.J. Tyler, F.H. Hambleton, S.A. Mitchell and
    J.A. Hockey, J. Phys. Chem., 73 (1969) 3947.
5   R.J. Peglar, F.H. Hambleton and J.A. Hockey, J. Catal., 20
    (1971) 309.
6   F.H. Hambleton and J.A. Hockey, J. Catal., 20 (1971) 321.
7   G.C. Armistead and J.A. Hockey, Trans. Faraday Soc., 63
    (1967) 2549.
8   M.L. Hair and W. Hertl, J. Phys. Chem., 73 (1969) 2372.
9   J. Kenawicz, P. Jones and J.A. Hockey, Trans. Faraday Soc.,
    67 (1971) 848.
10 P. Fejes, I. Kiricsi, I. Hannus, Á. Kiss and Gy. Schőbel,
    React. Kinet. Catal. Lett., 14 (1980) 481.
11 H.K. Beyer and I. Belenkaja, in B. Imelik et al. (Eds.),
    Catalysis by zeolites, Elsevier, Amsterdam, 1980, p. 203.

CHARACTERIZATION OF METAL AGGREGATES IN ZEOLITES

P. GALLEZOT and G. BERGERET
Institut de Recherches sur la Catalyse, 2 avenue Albert Einstein - 69626 - Villeurbanne Cédex - FRANCE

ABSTRACT

Metal aggregates trapped in porous network of zeolites have unusual atomic and electronic structure because of the size and of the environment effects. The different methods of characterization are examined critically, stress being laid on physical methods giving a direct and quantitative evaluation of size, location atomic structure and electronic structure of the aggregates.

INTRODUCTION

A detailed characterization of metal aggregates supported on zeolites is needed to interpret the catalytic properties of these materials which are involved in a number of catalytic processes of present or potential interest. Thus, in addition to their well known use in hydrocracking and hydroisomerization processes for converting low-priced petroleum cuts into fuel (ref. 1) metal zeolites have potential applications in the conversion of CO and H_2 into hydrocarbons by the Fischer-Tropsch synthesis (ref. 2). From the standpoint of fundamental researchs, metal zeolites are well suited materials for the study of the atomic and electronic structure of very small metal particles (aggregates, clusters). Indeed, particles can be obtained with sizes ranging from the dimension of an isolated atom to that of a cluster containing a few tens of atoms i.e. in a range where the intrinsic properties of a metal are no longer those of the bulk metal. Furthermore, because these particles are trapped in cages or pores they may exhibit a very narrow size distribution and a good stability toward sintering. Finally, the effects of the particle environment on the metal properties can be studied easily by modifying the Al/Si ratio, or the protonic acidity of the zeolite framework or by introducing multivalent cations or other additives.

The complete characterization of the metal aggregates comprises the determination of the metal dispersion (particle size and particle location) as well as the determination of the atomic and electronic structure of the particles. The aim of the forthcoming sections is to review critically the different methods of metal particles characterization, stress being laid on physical methods which give direct and quantitative information without any assumptions.

CHARACTERIZATION OF METAL LOCATION

The metal location in zeolites is even more important to determine than particle size, since it rules the metal-reagent accessibility and the support effect
on metal properties. The question is not only to settle if the particles are in
or out the zeolite crystals but also to determine what cages are occupied.
As far as diffusion and catalysis in zeolite pores are concerned there is also
the question as to whether the particles are uniformly distributed throughout
the zeolite crystals or concentrated in a given region e.g. in the zeolite outer
layers.

Electron microscopy techniques are the best suited to answer these questions.
Direct examination with the conventional transmission electron microscope (CTEM)
of the zeolite crystals dispersed on a carbon-coated grid can give clues to the
presence of metal on the external surface but the electron beam absorption in the
zeolite crystal hampers the detection of particles. This detection is best achieved on carbon film replica which extracts the metal particles from the external
surface. The detection of the particles in the zeolite crystals can only be carried out on transmission views through thin slabs (100-500 Å) of the zeolite crystal cut with an ultramicrotome equiped with a diamond knife. Examination by the
CTEM on slabs cut at different heights in a crystal and at different places of
a given slab give direct evidence for the presence of metal particles in the crystals and for the homogeneity of the particle location. Thus, it was shown that
20 Å Pd or Pt particles in Y zeolite are actually occluded in the zeolite bulk
although they are larger than the supercages (ref. 3). This type of investigation
is possible whenever the particles are large enough to be detected by the CTEM,
the limit depends upon both the resolution power of the CTEM and the contrast between the particles and the support. Modern microscopes allow the detection of platinum particles larger than 5 Å in the case of metals of the 2nd and 3rd transition row. Contrast is greatly improved by using dark field imaging techniques
(ref. 4).

Metal aggregates can also be detected, even if they are too small to be ima-
ged, by using the analytical capabilities of the scanning transmission electron
microscope (STEM). Thus, the VG, HB-5 STEM is equiped with an X-ray emission ana-
lyser and with an electron energy loss spectrometer which, because of the sharp-
ness of the beam, allow the analysis of minute amount of metal in a volume of
400 Å^2 projected area. These analytical tools are well adapted to check the homogeneity of the metal distribution, in the case of bimetallic catalysts they can
give the composition of the individual particles.

Surface enrichment in metal can also be detected by XPS, the ratio between a
XPS line of a metal and that of a support atom (e.g. Si) is then larger than the
ratio of these elements derived from the chemical composition. This technique is

useful to verify that no surface enrichment occurs in the course of metal loading in fresh catalysts or as a result of metal migration in used catalysts (ref. 5).

Crystal structure determination can be used in favorable cases to locate the metal atoms in the zeolite cages. In dehydrated zeolites, the cations occupy well defined sites in close interaction with the framework oxygen atoms. Upon reduction one can expect the following cases : (I) the atoms remain on similar sites but at a longer distance from the oxygen atom (II) they remain in the same type of cage but on random sites (III) they migrate out of the cage or along pores and agglomerate to form particles encaged in the zeolite (IV) they migrate out of the zeolite crystals. Cases (I) and (II) are favorable as far as crystal structure determination is concerned because the atoms still obey the periodicity of the zeolite lattice. Thus the crystal structure of partially reduced AgA (refs. 6, 7) and AgF (refs. 7, 8) (F = faujasite) zeolites have been determined from single crystal and powder X-ray diffraction data. Positively charged Ag_6 and Ag_3 silver clusters have been located on well defined sites in the sodalite cages of these zeolites. The crystal structure studies of $Pd_{12.5}Y$ (ref. 9) and $Pd_{13.7}Y$ (refs. 10, 11) zeolites have shown that after activation in O_2 the Pd cations occupy the SI' sites at x = y = z = 0.044 in strong interaction with the O(3) framework oxygen (table I). Upon reduction at 150°C the Pd atoms in $Pd_{13.7}Y$ occupy SI' sites at x = y = z = 0.082 at a longer distance from the O(3) atoms (2.74 Å) than before reduction (2.07 Å) (ref. 11). The AgY and $Pd_{13.7}Y$ samples are typical of case (I) since the reduced atoms are on well defined sites and still interact with the framework oxygen atoms. On the other hand the reduction of $Pd_{12.5}Y$ at 200°C is an example of case (II) because after reduction most of the atoms are in random positions in the same type of cage as before reduction. They have been localized using liquid scattering function corresponding to a model where the Pd atoms are distributed at any place within a 2.5 Å radius sphere in the sodalite cage. The fact that there are more Pd atoms occupying SI' sites in $Pd_{13.7}Y$ than in $Pd_{12.5}$ zeolite is probably due to a stronger interaction of the reduced species with the zeolite framework because the former zeolite is more acidic that the latter. The adsorption of C_6H_6 on the oxidized form of the $Pd_{13.7}Y$ zeolite also produces a limited reduction of type (I) since almost 6 atoms move from SI' (0.044) to SI' (0.082) (table I). The structure of a $Pt_{10}Y$ zeolite activated at 600°C and reduced at 300°C (ref. 12) is an other example of case (II) since the Pt atoms were found mostly on random sites in the sodalite cages (table I). In the PdY and PtY zeolites described in table 1 there are statistically only one or two atoms per sodalite cage so that the metal is truly atomically dispersed. However EPR (ref. 13) and XPS studies (refs. 14, 15) have shown that these Pd and Pt atoms are positively charged because they interact with the acid sites of the support, this is probably the reason why they do

TABLE 1

Distribution of palladium and platinum in the sodalite cages of PdY and PtY zeolites (refs. 9-12)

Samples and treatments ($^{\circ}$C, atmosphere)	Population of zeolite sites[a]				Interatomic distances (Å)	
	SI'(0.04)	*SI'(0.08)	SII'(0.17)	U(0.125)	SI'-03	*SI'-03
$Pd_{12.5}Na_{19.5}H_{11.5}Y$ 600, O_2	$10.6Pd^{2+}$				2.01	
$Pd_{12.5}Na_{19.5}H_{11.5}Y$ 600,O_2 - 200, H_2		$4Pd^{\delta+}$		$9Pd^0$		2.72
$Pd_{13.7}Na_{9.3}H_{19.2}Y$ 550, O_2	$12.0Pd^{2+}$				2.07	
$Pd_{13.7}Na_{9.3}H_{19.2}Y$ 550, O_2-150, H_2		$8.8Pd^{\delta+}$	$1.8Pd^{\delta+}$			2.74
$Pd_{13.7}Na_{9.3}H_{19.2}Y$ 550, O_2-25,C_6H_6	$4.1Pd^{2+}$	$5.7Pd^{\delta+}$			2.19	2.70
$Pt_{10}Na_{17}H_{19}Y$ 600, O_2	$9.8Pt^{2+}$				2.20	
$Pt_{10}Na_{17}H_{19}Y$ 600,O_2 - 300 H_2		$3(0.065)Pt^{\delta+}$		$8Pt^0$		2.54

[a] coordinates of the sites (x = y = z) are given in parentheses. (Sites U correspond to Pd or Pt atoms refined by liquid scattering function assuming that the atoms are uniformly distributed in a 2.5 Å radius sphere.

not chemisorb hydrogen (refs. 9, 12). Besides, they do not chemisorb (refs. 9, 12) or interact (ref. 16) with oxygen because they are inaccessible to the O_2 molecule. In case (III), where the encaged particles are formed at the expenses of migrating atoms, there is usually a very irregular occupancy of cages. Thus if 10 Å particles containing 20 M atoms are formed in a $M_{10}Y$ zeolite there are statistically only 1 supercage out of 16 occupied. In this case and whenever the particles are on the external surface (case III) the X-ray scattering by the metal atoms is independent from that of the zeolite lattice so the atoms cannot be located by crystal structure analysis. There are many indirect methods to check whether the metal particles are inside or outside the zeolite crystals and, in some favorable cases, to determine what type of cage are occupied. Any spectroscopic method capable of probing the interaction between metal atoms and adsorbed molecules can be used to determine the probable location of metal provided the size of the probe molecule is well choosen with respect to the cage aperture. However non-quantitative methods fail to give a representative picture of the true metal location.

PARTICLE SIZE MEASUREMENTS

Electron microscopy techniques are well suited to detect metal (see section 1) and therefore to measure particle sizes, however some important limitations should be kept in mind. Several hundreds of particles should be measured on different photographs to obtain a representative distribution of particle sizes. The main drawback is that particle size can be modified by the powerful electron beam in the CTEM and especially in the STEM. Electron beams are liable to produce high local temperature rises, reduction of cations, and recoil of loosely bound atoms. The combination of these effects may cause the migration of atoms -especially isolated atom or very small cluster and therefore lead to sintering.

X-ray line broadening has been used extensively as a method for determining crystallite size in metal supported catalysts. Unfortunately, it is of little help in the case of metal zeolite systems because crystallite sizes smaller than 20 Å cannot be measured. This limit holds in the most favorable cases e.g. when the zeolite contains at least 5 wt % of a 3 rd transition row element. The application of the method is usually restricted to the evaluation of metal particles on the external surface.

Small angle X-ray scattering (SAXS) is a valuable method to measure particle size of metal supported on zeolite. The main difficulty usually encountered in the study of metal supported catalysis by SAXS is that the heterogeneities in the support (pores, particles) give a strong SAXS from which that of the metal is difficult to separate. In the case of zeolites, the particles are large and the micropores are ordered so that the SAXS from the support is negligible with respect to that of the metal. Different parameters such as the radius of gyration the specific surface derived from Porod's law and the distribution of particle diameters can be derived from the analysis of the SAXS intensity curve (refs. 17, 18). The range of sizes which can be measured extends from a few to several hundred Ångströms and the sensitivity is sufficient to perform measurements with metal concentration as low as 0.5 - 1 wt % (ref. 18). Another important advantage is that SAXS integrates all the particles present in the samples so that the size parameters are representative of the actual metal distribution. The method has been used successfully to measure the platinum particles in a $Pt_{10}Y$ zeolites (ref. 12). The size distribution in $Pt_{10}Y$ treated at 300°C in O_2 and reduced at 300°C in H_2 was found in the 6-13 Å range with a maximum at 11 Å in agreement with CTEM measurements. In $P_{10}Y$ treated at 600°C in O_2 and reduced at 300°C in

H_2, where 75 % of the palladium atoms are isolated in the sodalite cages, the method fails to detect these atoms because they scatter X-rays over large Bragg angles. The specific surface deduced from the Porod's law therefore artifically too low and the equivalent particle diameter calculated from this surface is not representative of the real dispersion.

Particle sizes can also be measured by magnetic methods in the case of ferromagnetic metals. Thus the saturation magnetization method has been used to measure the reduction level and the particle size of nickel in a serie of NiY zeolites (ref. 19). The smallest particles (< 10 Å) can only be measured if the particle saturation is attained which requires measurements at 4,2 K. These results are perturbed when paramagnetic Ni^{2+} cations are present so that the method cannot be applied on partially reduced samples. Ferromagnetic resonance has been used to characterize the nickel particles formed inside or outside the NiNaY zeolites (ref. 20) and to characterize the presence of small homodispersed Ni particles obtained by reduction of NiX by H atom beams (ref. 21). Mössbauer spectroscopy has been used to characterize size and location of Fe particles in X and Y zeolite (ref. 22).

A novel, indirect method to evaluate particle size has been recently described by Fraissard et al. (ref. 23). They have shown that the study by NMR of the chemical shift of argon due to the collisions of the atoms with the cage walls and with the H_2-covered, encaged Pt particles, can be used to estimate the average number of metal atoms per particle. It was found that in a PtY zeolite this number was 8 Pt which corresponds to an average particle size of ca. 5 Å whereas the CTEM study indicated a 10 Å size average. This discrepancy could be due to the fact that a number of particles are not detected by CTEM.

The particle size can be determined from selective gas adsorption measurement on metals, however the stoichiometry of adsorption of H_2 and O_2 is still a matter of controversy and in the case of metal zeolite systems further complications arise. Firstly, the atoms in hidden positions are not accessible to O_2 or CO at 25°C. This is true for Pd and Pt atoms is the sodalite cages which do not chemisorb O_2 or CO because the molecules do not enter the cages. Secondly, metal can be in the form of charged clusters which do not behave like metal as far as chemisorption is concerned.

ATOMIC STRUCTURE OF PARTICLES

The atomic structure of metal particles can be determined by the Radial Electron Distribution method (RED) from X-ray diffraction data and by the Extended X-ray Absorption Fine structure (EXAFS) method from X-ray absorption data.

The RED is based on the Fourier transform of the intensities of X-ray scattered by any form of matter. This gives the distribution of all the interatomic vectors (ref. 24). The positions of the peaks gives the bond lengths and the surfaces of the peaks are proportionnal to the number of vectors and therefore to the coordination numbers around a given atom. Although the method has been applied with success to study supported metal catalysts with less than 1 wt % of platinum (ref. 24) larger metal concentrations improve the quality of the structural information. In the case of very small aggregates where most of the atoms are on the metal surface, the RED is sensitive to the surface structure and therefore to the modifications induced by the adsorbates. The structures of the 10 Å Pt particles have been studied after H_2, O_2, CO, H_2S adsorption (refs. 26-28) and in the course of C_6H_6 hydrogenation (29). Figure 1 gives the RED curves corresponding to 10 Å Pt particles in a PtCeY zeolite after adsorption of hydro-carbons under various conditions (ref. 30). When the particles are covered with H_2 (curve 2) the RED corresponds to the ordered f.c.c. packing and interatomic distances of bulk Pt whereas the structure of the naked particles (curve 1) can be described as a distorted and contracted f.c.c. structure. The RED obtained after adsorption on the naked particles of n-C_4H_{10} at 450 K (curve 3) of C_2H_4 at 300 K (curve 4) and the cracking of C_6H_6 at 600 K (curve 5) are very similar. With respect to the structure of the naked particles, the interatomic distances are slightly relaxed but the heights of the Pt-Pt peaks are smaller. These modi-fications are due to the formation of Pt-C bonds which improve the coordination of the surface atoms (relaxation of the distances) but at the same time produce a disorder among the Pt atoms (decrease of peak heights). Adsorption of H_2 at 300 K (curve 6) on the particles covered with cracked C_6H_6 produces a relaxation and a reorganisation of the structure because Pt-H bonds are formed, and treat-ment under H_2 at 600 K (curve 7) restores the initial, H_2 covered structure be-cause all the Pt-C bonds have been hydrogenolysed. These results as the preceding ones, indicate that the structure of the Pt aggregates depends upon the nature of the bonds formed between the metal atoms and the adsorbate molecules.

The EXAFS spectra obtained by X-ray absorption measurement on the high energy side of the absorption edges are due to the interference between the photoelec-trons ejected from the absorbing atom and the photoelectrons back-scattered from the neighbouring atoms (ref. 31). The Fourier transform of the EXAFS oscillations gives a radial distribution from which the interatomic distances are derived. The method has two important limitations (I). It is mainly sensitive to the nea-rest neighbours of the absorbing atom so that little informations can be obtained on the arrangement of atoms in the successive coordination spheres (II) the peaks on the Fourier transform are shifted downward with respect to the true interato-mic distances by 0.1 - 0.3 Å. The so-called phase shift problem is generally sol-ved by assuming that the phase-shift correction is the same than in a model com-pound (e.g. bulk metal). However, the validity of the phase shift transfer may be

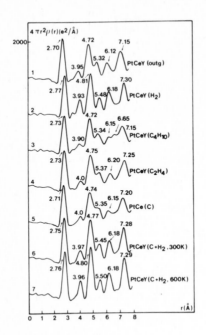

Figure 1 : Radial electron distribution of a PtCeY zeolite containing 10 Å particles covered with various adsorbates (from ref. 30), peaks correspond to the interatomic distances between the platinum atoms.
(1) PtCeY outgassed (2) H_2 at 300 K (3) n-C_4H_{10} at 450 K (4) C_2H_4 at 300 K (5) cracking of C_6H_6 at 600 K (6) regeneration under H_2 at 300 K (7) regeneration under H_2 at 600 K.

dubious when the electronic structure of the metal atoms is very different from that of bulk metal (e.g. in the case of very small charged clusters). The Pt-Pt distances in Pt particles smaller than 10 Å in the acidic PtCaY and PtHY zeolites are contracted with respect to the distances observed in PtNaY (ref. 32), the question is not settled whether these differences are due to a true contraction of the distances or to differences in the size or electron deficiency of the Pt aggregates. The EXAFS method provides informations (I) on the distances between the metal atoms and the support or adsorbate atoms whereas the RED method gives only the metal-metal distances (II) on the environment of individual atomic species. The RED and the EXAFS methods rely on data which can be obtained under various temperature and atmosphere conditions so that the particle structure can be determined during in situ experiments. The relaxation of the Pt-Pt distances when H_2 is adsorbed on the bare Pt particles in Y zeolites has been observed both by RED (refs. 29-33) and by EXAFS (ref. 34).

ELECTRONIC STRUCTURE OF METAL AGGREGATES

There is a corpus of theoretical works showing that the electronic structure
of metals is bound to change when particle sizes decrease below 10-15 Å. Furthermore, the environment of the aggregates induces additional perturbation of the
electronic structure. XPS has been the most used method to probe the electronic
structure of metals in zeolites. Thus the valence state of various metal has
been studied by Minachev et al. (refs. 5, 15). The XPS binding energies of platinum in well characterized PtY and PdY zeolite have been measured by Vedrine
et al. (ref. 14). The Pt 4 $f_{7/2}$ lines are shifted by + 0.7 eV with respect to
bulk metal in the case of 10 Å Pt particles and by 1.3 eV in the case of isolated Pt atoms. The platinum is electron deficient because of an electron transfer
from the metal atom to the electron acceptor sites of the support. Additional e-
vidence for this support effect were given by an EPR study of the charge transfer complexes formed in PtY and PdY zeolite (ref. 14). The main limitation of the
XPS method is that it only probes the metal on the external surface and in the
outer layers of the zeolite crystals which might not be representative of the
main fraction of the metal encaged deeper in the crystals. Thus in Mo - rich
PtMoY zeolites there is a surface enrichment in Mo and these species not bonded
to platinum are in a higher valence state than those involved in Pt-Mo associa-
tions (ref. 35).

X-ray absorption spectroscopy is another direct method of evaluating the e-
lectronic structure of metal. Thus Pt L_{III} edge spectroscopy has been used to
probe on a relative basis the charge density on 10 Å Pt particles as a function
of the acidity of the support, multivalent cations and adsorbates (refs. 16, 36).
The Pt LIII edge involves electronic transitions from 2 p level to the empty 6s
and 5d states of the valence band producing a so-called white line in the X-ray
absorption spectrum whose area is related to the number of unoccupied levels
in the valence band. The electron deficient or electrophilic character of the
10 Å particles follows the sequence Pt metal < PtNaY < PtNaHY < PtCeY < PtCaY ≃
PtHY,therefore it depends upon the Brönsted acidity and on the presence of multivalent cations. Furthermore the electrophilic character increases when electron
acceptor atoms are adsorbed (S) and decreases when electron donor molecules
(NH_3) are adsorbed.

Infrared spectroscopy studies of CO and NO adsorbed on metal have been used
extensively to probe the electronic properties of metals, the method has been
applied on PdY (ref. 13, 37) and PtY zeolites (ref. 38) to demonstrate the electron transfer from the metal to the support and between the metal and the adsorbates. Other physical methods have been used to probe the electronic properties
of metals in zeolites, a complete review is beyond the scope of this article. Although rapid progress is made in that field, it remains difficult so far to

decide wether the unusual electronic structure of metal aggregates is due to intrinsic size effects or to environment effects.

Chemical methods can, in certain cases, be best suited to evaluate the electronic properties of metal aggregates. Thus it has been shown that the ratio of the absorption coefficients of benzene and toluene can be derived from the kinetics of the competitive hydrogenation reaction of these hydrocarbons. The values of these ratios are more sensitive than XPS or X-ray absorption spectroscopy to evaluate the electron deficient character of the platinum aggregates (ref. 39).

CONCLUSION

Because of the smallness of the particles encaged in zeolites the characterization of metals is a difficult task. Several physical methods attain their limit of applicability when the particles size decrease below 10 Å, or a least, give results which are difficult to interpret. Therefore a precise characterization should be carried out with several methods. Moreover, these methods should be applied on the same material treated under similar conditions because the dispersion and the structure of the metal aggregates are highly sensitive to the thermal treatments and to the nature and pressure of the gas adsorbed. The characterization of the location, dispersion and structure of metal particles should be carried out on the freshly prepared catalysts and on the used catalysts because modifications are liable to occur during catalytic reaction. Ideally, the catalysts should be studied under working conditions by in situ experiments. In this respect X-ray methods are well suited and will probably expand in the forthcoming years thanks to the new powerful X-ray sources (synchrotron radiation) and new detectors available. Also the RED and the EXAFS methods are at present the sole techniques capable of determining the atomic structure of metal clusters. Electron microscopy techniques will also expand thanks to the addition of analytical tools and treatment chamber in the microscope column. The electronic structure of metal aggregates will remain difficult to determine because of the combination of   intrinsic size effects and of environment effects.

REFERENCES

1. A.P. Bolton in J.A. Rabo(Ed.), Zeolite Chemistry and Catalysis, American Chemical society, Washington, 1976, p. 714.
2. P.E. Jacobs in B. Imelik et al. (Ed.), Catalysis by Zeolites, Elsevier Amsterdam, 1980, p. 293.
3. P. Gallezot, I. Mutin, G. Dalmai-Imelik and B. Imelik, J. Microsc. Spectrosc. Electron, 1, 1976, 1.
4. M.J. Yacaman and J.M. Dominguez, J. Catal., 67, 1981, 475.
5. K.H. Minachev, G.V. Antoshin, E.S. Shpiro and Yu. Usifov, Proc. 6th Int. Cong. on Catalysis London 1976, Vol 2, the Chemical Society, London 1977, p. 621.

6. Y. Kim and K. Seff, J. Am. Chem. Soc., 99, 1977, 7055.
7. L.R. Gellens, W.J. Mortier and J.B. Uytterhoeven, Zeolites 1, 1981, 11.
8. L.R. Gellens, W.J. Mortier and J.B. Uytterhoeven, Zeolites 1, 1981, 85.
9. P. Gallezot and B. Imelik, Adv. Chem. Ser., 121, 1973, 66.
10. G. Bergeret, P. Gallezot and B. Imelik, J. Phys. Chem., 85, 1981, 411.
11. G. Bergeret and P. Gallezot, J. Phys. Chem., submitted for publication.
12. P. Gallezot, A. Alarcon-Diaz, J.A. Dalmon, A.J. Renouprez and B. Imelik,
    J. Catal., 39, 1975, 334.
13. C. Naccache, M. Primet and M.V. Mathieu, Adv. Chem. Ser., 121, 1973, 266.
14. J.C. Vedrine, M. Dufaux, C. Naccache and B. Imelik, J. Chem. Soc. Faraday
    Trans. 74, 1978, 440.
15. G.V. Anthoshin, E.S. Shpiro, O.D. Tkatchenko, J.B. Nikishenko, M.A. Ryashent-
    sver, V.I. Avaev and Kh. M. Minachev, Proc. 7th Int. Congr. Catalysis,
    Tokyo, 1980, Elsevier Amsterdam, 1981, p. 302.
16. P. Gallezot, R. Weber, R.A. Dalla Betta and M. Boudart, Z. Naturforsch, 34A,
    1979, 40.
17. A. Guinier, X-ray diffraction, W.H. Freeman, London, 1963, p. 319.
18. A. Renouprez, C. Hoang-Van and P.A. Compagnon, J. Catal., 34, 1974, 411.
19. M.F. Guilleux, D. Delafosse, G.A. Martin and J.A. Dalmon, J.C.S. Faraday I,
    75, 1979, 165.
20. P.A. Jacobs, H. Nijs, J. Verdonck, E. Derouane, J.P. Gilson, and A. Simoens,
    J.C.S. Faraday Trans. I, 75, 1979, 1196.
21. M. Che, M. Richard and D. Olivier, J.C.S. Faraday, Trans. I, 76, 1980, 1526.
22. E. Schmidt, W. Gunsser and J. Adolph, in Molecular Sieves II, American Chemi-
    cal Society Washington, 1977, p. 291.
23. L.C. de Menorval, T. Ito and J.P. Fraissard, J.C.S. Faraday Trans. I, 78,
    1982, 411.
24. P. Ratnasamy and A.J. Leonard, Catal. Rev-Sci. Eng., 6, 1972, 293.
25. P. Ratnasamy, A.J. Leonard and J.J. Fripiat, J. Catal., 29, 1973, 374.
26. P. Gallezot, A. Bienenstock and M. Boudart, Nouv. J. Chim., 2, 1978, 263.
27. P. Gallezot, Zeolites, in press.
28. P. Gallezot, in V.C. Reeds, (Ed.) Proc. 5th Int. Conf. Zeolites, Heyden, Lon-
    don, 1980, p. 364.
29. P. Gallezot and G. Bergeret, J. Catal., 72, 1981, 294.
30. P. Gallezot, J. Chim. Phys., 78, 1981, 881.
31. E.A. Stern, D.E. Sayers and F.W. Lytle, Phys. Rev., B11, 1975, 4836.
32. R.S. Weber, M. Boudart and P. Gallezot in J.B. Boudon (Ed.), Growth and Pro-
    perties of Metal Clusters, Elsevier, Amsterdam, 1980, p. 415.
33. P. Gallezot, Surface Sci., 106, 1981, 459.
34. B. Moraweck, G. Clugnet and A.J. Renouprez, Surface Sci., 81, 1979, 631.
35. Tran Manh Tri, J.P. Candy, P. Gallezot, J. Massardier, M. Primet, J. Vedrine
    and B. Imelik, J. Catal., submitted for publication.
36. Tran Mahn Tri, J. Massardier, P. Gallezot and B. Imelik in T. Seryama and
    K. Tanabe (eds). Proc. of the 7th Int. Cong. on Catalysis, Elsevier, Amster-
    dam 1981, p. 266.
37. G.D. Chukin, M.V. Lonaon, V. Kruglikov, D.A. Agievskii, B.V. Smirnov,
    A.L. Belozerov, V.D. Asriever, N.V. Goncharova, E.D. Rodchenko, O.D. Konoval-
    cherov and A.V. Agafonoy, in "Proc. 6th Int. Cong. on Catalysis, Vol 1.,
    Chemical Society London 1977, p. 621.
38. P. Gallezot, J. Datka, J. Massardier, M. Primet and B. Imelik in "Proc. 6th
    Int. Cong. on Catalysis Vol 2, Chemical Society, London, 1977, p. 696.
39. Tran Manh Tri, J. Massardier, P. Gallezot and B. Imelik, C.R. Acad. Soc.
    Paris, 293 II, 1981, 35.

NUCLEAR MAGNETIC RESONANCE STUDY OF XENON ADSORBED ON METAL-NaY ZEOLITES (SIZE OF METAL PARTICLES AND CHEMISORPTION).

J. FRAISSARD[1], T. ITO[2], L.C. de MENORVAL[1] and M.A. SPRINGUEL-HUET[1]

[1]Laboratoire de Chimie des Surfaces, associé au C.N.R.S. ERA 457/02, Université Pierre et Marie Curie, Tour 55, 4, Place Jussieu, 75230 Paris Cedex O5 (France).
[2]Research Institute for Catalysis, Hokkaïdo University, Sapporo 060, (Japon).

ABSTRACT

The NMR chemical shift of ^{129}Xe adsorbed on zeolites is very large and sensitive to the environment of xenon used as a probe. Among the results obtained, the most important show that the average number of atoms per particle of metal supported on zeolite is always below 10, whereas electron microscopy showed that the particle size is ca 10A. There therefore appear to be very small metal particles not detected by electron microscopy. This technique is also useful to study the adsorbate distribution and to follow step by step certain simple reactions at the surface of these metals.

INTRODUCTION

The central idea of this research was to find a non-reactive molecule, particularly sensitive to its environment and to collisions with other chemical species, which could serve as a probe for determining in a new way certain properties of solid catalysts, and, especially, zeolites. In addition this probe should be detectable by NMR since this technique is particularly suitable for the detection of electron perturbations in rapidly moving molecules.

After many tests on several molecules such as $^{15}N_2$, ^{3}He... we have chosen xenon. The inert gas Xe is an ideal probe because it is monoatomic and has a spherically symmetrical electron distribution. Any distorsion of this latter is transmitted directly to the Xe nucleus and will affect the NMR chemical shift. The ^{129}Xe isotope to be studied has no nuclear quadrupole moment and the natural abundance in Xe is 26.44 %. The NMR sensitivity of ^{129}Xe relative to an equal number of protons at constant magnetic field is 2.12×10^{-2}. Thus the minimum detectable number of Xe atoms in the NMR probe coil is $\sim 1.10^{20}$. This number is sufficient for precise measurement on zeolite-Xe systems with a reasonable signal-to-noise ratio.

The NMR chemical shift of xenon adsorbed on zeolites is very large (ref.1) . It depends on several factors : collisions between the xenon atoms and the walls of the supercage, the cations or other xenon atoms, the number and nature of the compensating cations, local electric fields, etc. Xenon can therefore be considered as a probe which is very sensitive to its environment.

We summarize here a certain number of results which demonstrate the interest of this new technique in : a) the study of the hydrogen or oxygen distribution on very small metal particles supported on zeolites. b) The determination of the number of these caged particles and of the mean number of metal atoms which each one contains. c) The study of the adsorption of various gases (CO, C_2H_4, H_2, O_2...) on these particles, simultaneous adsorption, reaction in the adsorbed phase, etc.

EXPERIMENTAL

Materials

The Pt_x-NaY and Ir_x-NaY samples were prepared by the method of Gallezot *et al* (ref.2) from Linde SK-40 zeolite, the cations of which had been partially exchanged by $[Pt(NH_3)_4]^{2+}$ or $[Ir(NH_3)_5Cl]^{2+}$ respectively.

x denotes the mean number of atoms of metal M (Pt, Ir) per particles, this number being determined at the end of the experiment by the method proposed.

Electron microscopy shows that the particule size is *ca* 10 Å whatever the sample ; therefore all the M atoms are not on the surface. The M_x-NaY zeolite studied was treated at 400°C and 10^{-5} torr. Xenon was then adsorbed, either on this treated sample or on similar samples on which a gas G (H_2, O_2,CO, C_2H_4) had been preadsorbed at room temperature. We would remark at this point that the results are exactly the same if adsorption of the gas G takes place in the presence of xenon. The latter samples will be denoted M_x- βG-NaY where β G is the ratio of the number of chemisorbed G atoms or molecules (if chemisorption is non dissociative) to the total number of M atoms.

Apparatus and procedure

NMR absorption of ^{129}Xe was observed at room temperature by using a Bruker pulse Fourier-transform spectrometer at a frequency of 24.9 MHz and a magnetic field of 21.14 kG stabilized by proton NMR. The number of accumulations necessary to obtain an acceptable signal-to-noise ratio depends, of course, on the Xe concentration. However, with a radio-frequency pulse repeat time of 0.5 s it was never less than 1000. The reference signal of ^{129}Xe is taken as that of Xe gas extrapolated to zero pressure according to Jameson's equation (5). All resonance signals of ^{129}Xe adsorbed on zeolites were shifted to higher frequency relative to the reference. This is defined as the positive direction in this paper.

RESULTS

Theoretical chemical shift of ^{129}Xe adsorbed on a zeolite

The chemical shift of xenon adsorbed in a zeolite is given by the sum of three terms characteristic of each of the effects which a gas can undergo there (ref. 1,3).

$$\delta(Xe) = \delta_O + \delta_S + \delta_E + \int_0^{\rho Xe} \delta(Xe-Xe)\, d\rho_{Xe} \qquad (1)$$

δ_O is the reference. δ_E is due to the electric field created by the cations. This term is negligible in the case of NaY or HY. $\int_O^{\rho Xe} \delta$ (Xe-Xe).dρ_{Xe} where ρ_{Xe} is the density of xenon adsorbed in the cavities, corresponds to the increase in shift caused by Xe-Xe collisions.

$$\delta \text{ (Xe-Xe)} = 0.40 \pm 0.02 \text{ ppm.amagat}^{-1} \qquad (2)$$

δ_S due to collisions between xenon and cage walls, is of the form

$$\delta_S = \delta\text{(Xe-zeolite)}.\rho_S \qquad (3)$$

δ (Xe-zeolite) is characteristic of collisions between the xenon and the silica-alumina surface of cages.

$$\delta \text{ (Xe-zeolite)} = 1.15 \pm 0.004 \text{ ppm.amagat}^{-1} \qquad (4)$$

ρ_S, corresponding to a density, depends only on the cage structure.

The chemical shift of the signal characteristic of xenon adsorbed in the supercages of NaY or HY zeolite (the Xe atom is too big to enter the sodalite cavities) is given by equation (1) where $\delta_E = 0$ and $\delta_S = 58 \pm 2$ ppm (ref.3). It varies linearly from 58 to 110 ppm when the number of xenon atoms per supercage goes from 0 to 3 (see fig.2). This chemical shift is denoted by δ_{NaY}.

Let us assume now that distributed in various NaY zeolite supercages there are solid particles with chemically different surfaces S_i. The index i denotes a particle type of population n_i, amongst the p types which exist ($0 < i < p$, see fig.1). Corresponding to each of these populations there is a shift δ_{S_i} characteristic of the Xe-S_i collision, being the product of two terms :

$$\delta_{S_i} = \begin{bmatrix} \text{term for the chemical} \\ \text{nature of } S_i \end{bmatrix} \times \begin{bmatrix} \text{term for the number} \\ \text{of Xe-}S_i \text{ collisions} \end{bmatrix}$$

If the lifetime of xenon on each surface S_i is long (from an NMR point of view) the spectrum of the adsorbed xenon should theoretically contain as many components δ_i as there are target types, the intensity of each being proportional to the number n_i of targets i of the same type ; the total being $p+1$ since the component (shift δ_{NaY}) due to xenon striking the supercage surfaces and the other Xe must also be taken into account. This case can be obtained by recording the spectrum at sufficiently low temperature. But then the components are very broad and almost undetectable.

In the opposite case where the xenon has a very short lifetime at each adsorption site and can, moreover, diffuse rapidly across several supercages contained in the same cristallite of the zeolite (for example, very rapid diffusion from point A to point K in figure 1) all the above signals coalesce. The spectrum consists of only one component whose position depends on the values of δ_i, each weighted by the probability α_i of Xe-S_i collision :

$$\delta = \Sigma \alpha_i \delta_i + \alpha_{NaY} \delta_{NaY}$$

with $$\qquad (5)$$

$$\Sigma \alpha_i + \alpha_{NaY} = 1$$

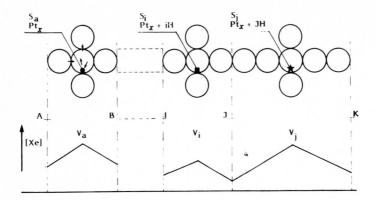

Fig.1. Solid particles distributed in zeolite supercages.Different xenon-target collisions and gradient of xenon concentration.

One can finally find a case intermediate between the two previous ones. Assume that the solid targets S_i, S_j.... are far fewer than the supercages and fairly distant from one another,*i.e.* separated by at least two or three empty supercages (see fig. 1). Let us assume also that these targets are the strongest xenon adsorption sites, that is, that they create a gradient of gas concentration, this being a maximum at S_i and decreasing rapidly along a few neighbouring supercages (see fig.1). We have therefore theoretical surfaces of lower concentration defining volumes V_i including a few super-cages around each target S_i. In each of them there can be rapid exchange between Xe atoms striking S_i, the cage walls and the other Xe. But because of the high concentration gradient, fast diffusion from one volume V_i to another volume V_j (see fig.1) does not occur. The NMR spectrum must therefore contain p components corresponding to each type of target, with shift δ_i, such that :

$$\delta_i = \lambda_i \, \delta_{S_i} + \mu_i \, \delta_{NaY} \tag{6}$$

λ_i and μ_i are the probabilities of xenon collision with S_i and NaY, respectively, in each volume V_i.

The case of Mx- βG-NaY samples.

Let us consider now a catalyst M_x - βG-NaY containing n metal particules.g^{-1} of sample (each containing on average x atoms of metal M) with a total amount β of adsorbed gas G (defined in materials). There is an atom G (or molecule) distribution over the totality of the metal, the number i of G atoms (or molecules) per particle being possibly different from one particle to another. Now, the chemical nature of a metal particle change with the number i of G atoms per particle. For each number i there is therefore a corresponding term $\delta_i = \delta_{Mx} + iG$ characteristic of the collision between Xe and the particle Mx + iG. Furthermore, the adsorption isotherms of xenon show that the presence of metals in NaY increases the adsorption of this gas, compared to pure NaY, at least for xenon pressures below about 400 torr. It is observed then that xenon does not diffuse rapidly across numerous supercages. Such a sample resembles the third case examined above : rapid exchange of xenon adsorption sites, but within restricted volumes containing a few supercages around the metal particles. The xenon spectrum must therefore contain as many components δ_i as there are particles distingui- shed by the number i of atoms G adsorbed ; the intensities of the components should be proportional to the number of particles in each group and the chemical shift will be given by equation (7) analogous to (6) :

$$\delta_i = \lambda_i \, \delta_{Mx + iG} + \mu_i \, \delta_{NaY} \tag{7}$$

From the components actually detected it is therefore theoretically possible to determine : firstly, the distribution of the adsorbate G on the metal particles for a total adsorption β ; secondly, the number n of particles in the sample and, consequen- tly, knowing the total metal content, the mean number x of M atoms per particle.

The above analysis is independent of the nature of the adsorbed gas. One can therefore treat an adsorbate containing several gases in the same way. We would point out also that, whatever the gas G, the number of NMR components is always very small. This facilitates the analysis of spectra and proves that the distribution of adsorbates on these small particles is always very simple. Let us now examine a few examples.

Samples Mx-βH-NaY. Mean number of atoms in small metallic particles.

Pt_x-βH-NaY (ref.4). The spectrum of xenon adsorbed on the Pt-NaY sample contains one line (denoted by a) whose chemical shift, δ_a (see fig.2) is always much greater than δ_{NaY} whatever the Xe pressure. For example, for an average of one Xe atom per super- cage $\delta_{NaY} = 78$ ppm and $\delta_a = 270$ ppm. The spectrum of xenon adsorbed on a sample Pt-βH-NaY, with a very small amount of pre-chemisorbed hydrogen (βH $<<$ 1/15) has two lines. The first is none other than line a , situated exactly where it is for βH = 0 at the appropriate Xe pressure. The position δ_b of the second line b is intermediate between δ_{NaY} and δ_a for the same Xe pressure (see fig.2). For example, for an average of one Xe atom per supercage $\delta_b = 163$ ppm : like δ_a, δ_b is also indepen- dent of βH(= 0) as long as βH is very small.

When βH increases, although still small, the magnitude of line a decreases in parallel with that of line b . The latter is greatest when a disappears. We write β_1H as the value of βH corresponding to this maximum in b. The variations of the signal strengths I_a and I_b with βH are non-linear. In contrast $I_b/(I_a + I_b)$ is linearly dependent on βH up to $\beta_1 H$ and is independent of the xenon pressure. The relaxation times T_{1a} and T_{1b} are the same and together with the linewidths ΔH_a and ΔH_b, are independent of βH. Figure 2 shows that the difference $\Delta\delta = \delta_a - \delta_b$ increases as [Xe] decreases. ΔH_a and ΔH_b also increase as [Xe] decreases but much less than $\Delta\delta$. The components a and b are therefore much better separated at low [Xe] values. However in this case the sensitivity is low. It is therefore necessary to reach a compromise between the resolution of components a and b and their ease of detection.

When βH increases beyond $\beta_1 H$, for a given Xe pressure, signal b diminishes and broadens with a shoulder towards low δ values. This shoulder corresponds to a new line c, chemical shift δ_c, which increases with βH at the expense of b. This component c is at a maximum when b disappears, which occurs when $\beta H = \beta_2 = 2\beta_1 H$. Since the resolution of the components b and c is poor except for $\beta H \sim \beta_1 H$ (line b) and $\beta H \sim \beta_2 H$ (line c), it is not possible to state that δ_c is independent of βH, but its variation, if any must be small. Of course, the value of δ_c depends on [Xe] just as δ_a and δ_b do (see fig. 2) ; the above results are reproducible for all values of [Xe]. In particular $\beta_2 H$ is always equal to $2\beta_1 H$.

In the same way one obtains a line d with a chemical shift δ_d by evolution of line c with βH ($> \beta_2 H$) just as this was obtained from line b. This line is strongest for $\beta_3 H \sim 3\beta_1 H$ but the difference ($\delta_c - \delta_d$) is too small for us to affirm that it is not a continuous shift of line c with increasing βH ($> \beta_2 H$). Moreover, even when it is strongest, line d is still broadened downfield at the base. This broadening is not negligible and could correspond to a residue of the c component.

Chemisorption of hydrogen on platinum is rapid at ambient temperature. However, the high-power NMR spectrometer used enabled us to record a few spectra before the H_2 chemisorption equilibrium was reached. For this purpose we introduced into the sample tube an exact amount of hydrogen , βH, in the presence of xenon in order to follow the development of the spectrum of the inert gas adsorbed. When $\beta H > \beta_1 H$ the first spectrum recorded after the introduction of H_2 contains the a and b components and a very broad shoulder on the latter, toward low δ , which could correspond to a set of lines c, d, etc. This shoulder decreases with time and after a few minutes there remain only lines a and b , whose relative magnitudes depend on the value of βH. When $\beta H = \beta_2 H$, the first spectrum recorded includes a fairly strong line c with a shoulder towards low δ and weak a and b lines. The shoulder and lines a and b have disappeared in the spectrum recorded a few minutes later, whereas line c has reached its maximum intensity.

We have shown that the chemical shift of each of these components a, b and c is given by equation (7) where M = Pt and i represents the number of chemisorbed H atoms per particle. The fact that the spectrum of xenon adsorbed on Pt-βH-NaY contains at the most only two components (as long as βH is small) proves that the hydrogen distribution over all the particles is very regular whatever βH. More precisely, we have shown that the numbers i and i' of two consecutive signals are such that $i' = i + 2$. This means that the signals a, b and c represent particles bearing 0,2 and 4 H atoms, respectively. One sees immediately the interest of this distribution for determining the number n of particles in a sample. This number is equal to the number of hydrogen molecules corresponding to $\beta_1 H$. Knowing the platinum concentration of the sample one deduces the mean number x of Pt atoms per particle. Table I presents the results obtained for the four samples examined and shows that x is between 7.8 and 3.7.

TABLE 1

Distribution of the platinum supported on zeolite.

| Sample | % by weight | Particule size from electron microscopy | Number of Pt particules relative to | | Number x of Pt atom per particule |
			one gram $\pm 0.02 . 10^{20}$	$3.76 . 10^{20}$ cages $\pm 0.02 . 10^{20}$	
Pt_8-NaY	14.0	10 Å	0.56	0.64	7.8
Pt_6-NaY	5.03	"	0.25	0.26	6.1
Pt_5-NaY	7.84	"	0.45	0.49	5.4
Pt_4-NaY	14.0	"	1.175	1.34	3.7

$Ir_x - \beta H$-NaY. The case of supported iridium is very similar to that of platinum (see fig.3), but the result is even more surprising because there are on average 2.5 Ir atoms per particle ! This number is however in agreement with the work of Mc Vicker et al (ref.6) who suggested on the basis of chemisorption measurements that there were many isolated Ir atoms in these samples and the possibility (which we have confirmed) that these atoms chemisorb 2 hydrogens. We would point out however that the shifts of lines b and c diminish slightly when β goes from o to $\beta_1 H$ and from $\beta_1 H$ to $\beta_2 H$ respectively. In our opinion this is due to the fact that the number of particles is 1/3 that of the supercages. In this case certain particles can be in the same cage or in adjacent cages. Each NMR signal is then due to the coalescence of more than two components. The value $\beta_1 H$ must only be determined therefore on the basis of the disappearance of line a.

Our result is quite different from that obtained by electron microscopy, which shows that the size of the particles detected is ca 10 Å, $i.e$, corresponding to ca 55 atoms, assuming a cubo-octahedral or icosahedral cluster (ref.7).

Comparison of the two results indicates that many particles contain very few atoms and cannot therefore be detected by electron microscopy. It should be pointed

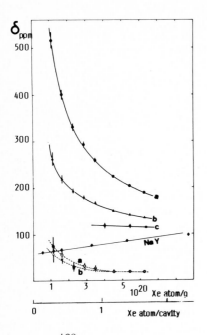

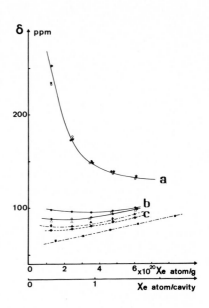

Fig.2 : ^{129}Xe n.m.r. chemical shift (full line) and line width (dotted line) as a fonction of the concentration of hydrogen adsorbed on Pt$_8$-NaY.

  NaY ; a : Pt-NaY; b : Pt-NaY+2H per Pt particle ; c : Pt-NaY+4H per particle.

Fig. 3 : ^{129}Xe n.m.r. chemical shift as a fonction of the concentration of hydrogen adsorbed on Ir$_{2.5}$-NaY.
a : Ir-NaY ; b: Ir-NaY + 2H per particle ; c : Ir-NaY + 4H per particle.

out, however, that the electron microscopy result and ours are not incompatible if it is assumed that, because of strong platinum-zeolite interactions, the small metal particles can be developed in the solid mainly in two dimensions.

  We would point out that, for electron microscopy characterisation of small metal particles, the imaging mode which we have used is the standard bright field (*i.e.* images which are formed using a beam which does not undergo diffraction either in the particle or in the support). Yacaman and co-workers (ref.8,9) have shown that this technique is extremely limited and proved that only the little-used dark-field techniques (*i.e* images which are formed using beams diffracted in either the particle or the support) allow the observation of small metal clusters down to a diameter of 5 Å.

Samples Pt_x-β(O)-NaY

The dependence on β(O) of the spectrum of xenon adsorbed on Pt_x-β(O)-NaY samples is entirely analogous to that found for H_2 : as long as β(O) is less than $β_1$(O) each platinum particle bears zero or two oxygens. For moderate xenon pressures, the chemical shifts of analogous signals corresponding to the chemisorption of H_2 or O_2 are different. For example, in the case of Pt_8-NaY, at a xenon pressure of 200 torrs, $δ_a$ = 290 ppm, $δ_b$ (2H) = 178 ppm and $δ_b$ (2O) = 165 ppm.

Sample Mx-β(CO) NaY

The chemisorption of CO at ambient temperature is different from those above. Whatever the value of β(CO) between 0 and 1/2, the spectrum of xenon adsorbed on Pt_x-β(CO)-NaY, for example, consists always of two signals : the first a corresponds to bare particules ; the second b , where $δ_b$(CO) $∿$ $δ_{NaY}$, is associated with "CO-saturated" particles where the stoichiometry of maximum chemisorption on these small particles is one CO for two Pt. Consequently, the first molecules introduced into the sample saturate the first platinum particles encountered in each zeolite crystallite. The number of these saturated particles increases with β(CO) at the expense of the bare particles. Figure 4 shows the variation of the relative intensities $I_a/(I_a + I_b)$ and $I_b/(I_a + I_b)$ with β(CO) when all the platinum is inside the supercages.

Conversely , at 300-400°C, the CO distribution is similar to that found for H_2 and O_2.

Application to the determination of the quantity of metal distributed on the external surface. When subjected to certain treatments the metal can partially migrate from the supercages to the external surface of the zeolite crystallites. During CO adsorption these external particles are saturated first. But the xenon spectrum concerns only the inside of the supercages. It changes therefore only when the internal particles begin to chemisorb CO, that is after saturation of all the external particles. Figure 4 displays the variation of the strengths of signals a and b with β(CO) for a sample of which 18% of the atoms accessible to CO are on the external surface of the crystallites. It is also possible to saturate the external particles with CO then chemisorb H_2 on the internal particles in order to count the latter.

Chemical reaction in the adsorbed phase : dehydrogenation of ethylene.

Our technique can be used to follow a chemical reaction in the adsorbed phase. As an example we give in figure 5 the case of C_2H_4 adsorbed on Pt_4-NaY containing $n = 1.02 \times 10^{20}$ particules.g^{-1}. In spectrum 1, line a of intensity I_{1a} is associated with bare particles. After adsorption of $n/3$ molecules of C_2H_4 at ambient temperature, spectrum 2 contains in addition to line a of intensity $I_{2a} ∿ 1/2\ I_{1a}$, two equally strong signals b and c : $I_{2b} ∿ I_{2c} ∿ 1/2\ I_{2a}$. These two last signals show that there are two types of "covered" metallic particles differentiated either by the amount of ethylene adsorbed (for example one molecule for line *b* ; two molecules for line *c*) or by the nature of the adsorbate consequent upon a change in the chemisorbed ethylene. In fact, it is the second explanation which is the good one. We have shown that line *b* corresponds to the chemisorption of two hydrogens per particle and line *c* to the adsorption of acetylene. It is easy to deduce then from this experiment that for low coverage at 25°C, chemisorbed ethylene decomposes immediately into a dehydrogenated complex and 2 hydrogens atoms, and that the latter migrate on the bare particles.

The small value of δ_{2c} suggests that the dehydrogenated complex occupies a large fraction of the surface of each particle and, therefore, that each of the two carbon atoms is attached to at least one Pt atom. Finally, analysis of the signal strengths indicates that not all the ethylene introduced is adsorbed on the metal.

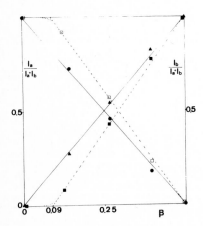

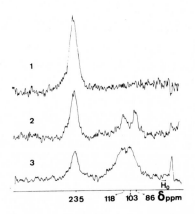

Fig. 4. ^{129}Xe n.m.r. chemical shift as a fonction of the concentration of CO adsorbed on Pt-NaY ; full line : all the platinum is inside the supercages ; dotted line : some platinum particles are on the external surface.

Fig. 5. ^{129}Xe n.m.r. chemical shift of xenon adsorbed on : 1 : Pt$_4$-NaY ; 2 : Pt$_4$-NaY + C$_2$H$_4$ at 25°C ; 3 : Pt$_4$-NaY + C$_2$H$_4$ at 60°C.

After heating in a sealed tube at 60°C, the sample gives spectrum 3 consisting of 3 lines : a) the intensity I_{3a} of line a is a third that of the initial I_{1a}. A third of the particles are therefore still bare. b) I_{3b}, I_{3c} et I_{3a} are roughly equal. This shows that the ethylene retained by the support at 25°C has migrated on the metal at 60°C. c) Signal b corresponds to particles carrying two hydrogens. The chemical shift of c has increased from 86 to 103 ppm. It is now close to the shift of particles bearing one CO. It corresponds to a coverage of the particle by the dehydrogenated complex smaller than in the case of spectrum 2. It seems reasonnable to suggest then that after heating at 60°C there is a reorientation of the complex ; the c-c bond which was initially parallel to the Pt-Pt bonds is now oriented perpendicularly to this bond to give a vinylidene group :

These results confirm very easily those (partial, since the H atoms were not located) of Somorjaï et al (ref. 10,11) for the (111) face of platinum, obtained by means of LEED.

CONCLUSION

The "xenon method" which we have developed is the only one presently able to detect zeolite supported metal particles containing very few atoms. In the study of chemisorption it allows us to work on real catalysts, at very low coverage but nevertheless with high sensitivity. It is moreover relatively specific since it is possible to count the different chemisorbed phases on the metals and often to establish the nature of each of them.

REFERENCES

1 T.Ito and J. Fraissard, Proceedings of the International Conference on Zeolites, Heyden, London, 1981, 510 pp.
2 P. Gallezot, A. Alarcon-Diaz, J.A. Dalmon, A.J. Renouprez and B. Imelik, J. Catal. 39 (1975) 334.
3 T. Ito and J. Fraissard, J. Chem. Phys. (in press).
4 L.C. de Menorval, T.Ito and J. Fraissard, J. Chem. Soc., Faraday Trans. I, 78 (1982) 403-410.
5 A.K. Jameson, C.J. Jameson and H.S. Gutowsky, J. Chem. Phys. 59 (1973) n°8, 4540
6 G.B. Mc Vicker, R.T.K. Baker, R.L. Garten and E.L. Kugler, J. of Catal., 65 (1980), 207-220.
7 M.B. Gordon, F. Cyrot-Lackmann and M.C. Desjonqueres, Surf. Sci., 68 (1977) 359.
8 M.J. Yacaman and J.M. Dominguez, J. Catal., 64 (1980) 1.
9 J.M. Dominguez and M.J. Yacaman, Growth and Properties of Metal Clusters, Elsevier, Amsterdam, 1980, 225 pp and 493 pp.
10 L.L. Kesmodel, R.C. Baetzold and G.A. Somorjaï, Surf. Sci., 66 (1977) 299.
11 L.L. Kesmodel, L.H. Dubois and G.A. Somorjaï, J. Chem. Phys. 70 (1979) 2180

CHARACTERIZATION OF METAL AGGREGATES IN ZEOLITES

F. SCHMIDT

Institut für Physikalische Chemie, Universität Hamburg,

D-2000 Hamburg 13, Laufgraben 24, West Germany

ABSTRACT

Transition metal aggregates encapsulated in the pore system of zeolites have been characterized by electron microscopy, Mößbauer spectroscopy and magnetic measurements. Even $Fe(0)$ and $Co(0)$ particles ($v^{1/3} \sim 5$ Å) exhibit the magnetic properties of the bulk material between approx. 5 K and 300 K. A strong coupling between the lattice vibrations of the metal and the matrix, but no electronic particle-support interaction was found.

INTRODUCTION

Microcrystals of transition metals supported on oxides such as SiO_2, Al_2O_3 or zeolites have widely been used as catalysts [1-4]. The first problem in order to characterize a supported catalyst is the analysis of the chemical composition and the determination of the various phases, their corresponding amount and volume. Furthermore, if the chemical composition, the structure and the volume are known, the true surface of ideal crystals can be determined from the known statistics of surface atoms [5]. To obtain the accessible as well as the catalytic active surface from chemisorption measurements the stiochiometry of a testing reaction must be known. In many practical cases this is obviously very difficult.

Particles, containing less than 1000 atoms, consist of more than 50% surface atoms. Thus, the knowledge of the state of the atoms directly in the surface of the microcrystals plays an important role for the understanding of a reaction in heterogeneous catalysis. However, geometry, lattice vibrations, diffusion, electronic and magnetic properties of these atoms may be quite different from the atoms in bulk ideal crystals [1].

Very many papers and numerous review articles [6-11] have been published on the structure and the properties of Pt- and Pd-particles supported on zeolites. The kinetics of formation, the physical and the catalytic properties of Ni clusters encapsulated in zeolites have been intensively studied, also [4,8,12-21]. Earlier literature is cited in these recent papers. However, iron microcrystals supported on zeolites have not been studied in greater detail, in spite of their very promising catalytic properties. In this study, we want to characterize aggregates of ferromagnetic metals - mainly iron - that are isolated in the pore system of zeolites. These investigations will have three aspects: First, from the theoretical point of view it is very interesting to know how many atoms are necessary to form a metal with solid state properties. This may depend on the particular property that is being considered [22].

Furthermore, the various influences on the formation and stability of microclusters may be of interest in order to prepare catalysts with definite properties. In this case, the prerequisite is to have reliable methods for the characterization of metal aggregates.

Finally, it is of basic importance to find methods for the characterization of a catalyst, which may be applied to 'in situ' studies of a supported metal during a catalytic process.

The main methods which have been used to characterize transition metals on zeolites are X-ray and electron diffraction, ESCA, EXAFS, Mößbauer spectroscopy (MES) and magnetic measurements. The last two methods are especially suited for making 'in situ' measurements. In this case, however, MES is restricted only to iron and the advantage of magnetic studies is limited to the investigation of ferromagnetic or superparamagnetic phases.

In order to compare experimental results of cluster properties with theoretical calculations, it is necessary to prepare uncharged microcrystals of definite structure, size and its corresponding narrow distribution, free of impurities. Until today it is impossible to produce such isolated particles, containing more than approx. 20 atoms. It could be shown, however, that zeolites are suitable for stabilizing such clusters [6-24]. Using this matrix, the knowledge of the particle-support interaction becomes of basic importance, if one wants to compare the experimental results with model calculations of a particular physical property [17,25].

PREPARATION

Zeolites loaded with transition metal ions have been prepared by exchanging stoichiometrically the sodium ions. For this purpose, nitrate or sulphate solutions of the divalent metal ions were used [14,26]. After dehydration, the reduction was performed by using various reducing agents such as H_2 - with and without UV-irradiation -, NH_3, Na- or K-vapour This was done to keep the reduction temperature as low as possible and the degree of reduction as high as possible.

Lattice and Lattice Dynamics of Metal Aggregates on Zeolites

In many cases the chemical composition, the amount and the volume of the metal can be determined by magnetic methods because many active catalysts contain superparamagnetic metals, oxides or carbides. An example for the thermomagnetic analysis (TMA) is given by Sancier et al. [27].

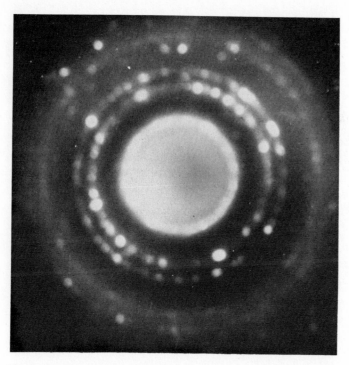

Fig. 1. Photograph showing electron diffraction image of Gd(0) particles in Y-zeolite.

Due to the fact that the lattice of the microcrystals does not need to be identical with that of the bulk material, experimental investigations are necessary. Next to the analysis of the X-ray interference function in case of very small clusters and Debye-Scherrer-plots - in case of larger particles - electronmicroscopy has been applied to study the lattice of microcrystals. In Fig. 1 the electron diffraction pattern of gadolinium clusters encapsulated in the cavities of Y-zeolite is shown. The preparation and the magnetic properties of these clusters have been described elsewhere [28]. These samples have been embedded in spurr, to prevent oxygen as well as water vapour from coming in contact with the previously reduced samples. After the sample had been cut with an ultramicrotom some of the smallest zeolite crystallites were still completely incased by the spurr. Therefore, in these crystallites the gadolinium particles remained in the metallic state. These particles produced the pattern shown in Fig.1. The d-values calculated from these experiments are in agreement with the values of bulk metal within the experimental error. Therefore, the structure of the clusters seems to be the same compared to the bulk material in spite of the fact, that the particle diameters are extremely small. The particle size was determined from the darkfield image which was produced by using a metal reflex (see Fig. 2). The size of the particles is approx. 1.3 nm.

These experiments allow not only the determination of the size and the structure of the particles but also provide some information concerning the geometrical arrangement of the metal aggregates on the zeolite support [25,29]. The particle arrangement in Fig. 2 agrees well with the distances and the angles between the supercages in the zeolite framework.

This interaction between the lattice of the metal particle and the support should have some influences upon the parameters of the ME spectrum.

The ME spectroscopy provides direct information concerning the type of state of the metal atoms in highly dispersed systems. Microcrystals differ from bulk material in the high amount of surface atoms. These atoms are influenced by the surroundings, which have a totally different structure especially in case of the metals supported on zeolites. The coupling which takes place, may effect the interface chemically, it may produce long range order

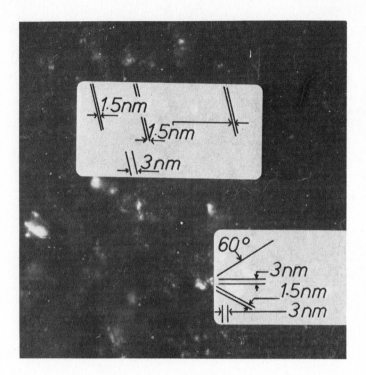

Fig. 2. Darkfield image of Gd(0) particles in Y-zeolite.

and it may generally influence the phonon spectrum.

Investigations by Van Wieringen have shown [30] that a minimum of $4 \cdot 10^5$ iron atoms per cluster is necessary to absorb the recoil momentum of the nuclear decay. On the other hand, we can conclude from magnetic measurements (see below), that in the case of iron clusters supported on A-zeolites, most of the particles contain less than 175 atoms per cluster. This is in agreement with the observation of a superparamagnetic doublet in the ME spectrum of the same samples [26]. From the fact, that these small clusters show a ME at all, we must conlcude, that there must exist a strong interaction between the lattices of the metal aggregates and the zeolite.

The metal-zeolite systems prepared in the manner described above, usually contain a very large amount of extremely small clusters and only a small amount of larger particles (see below).

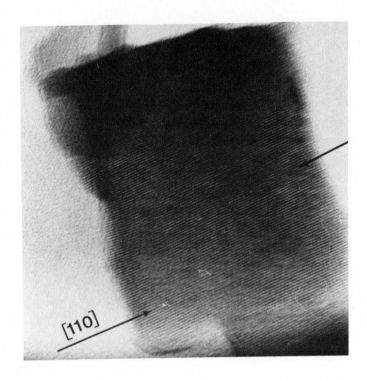

Fig. 3. Photograph showing an enlarged supercage (top of the
        arrow) in NaY-zeolite.

The large particles are located in a locally destroyed structure
of the matrix [11,21]. An enlarged cage of this type is shown
(see Fig. 3). The particles which fit in this kind of cavity or
which may be located on the external surface of the zeolite crystallites exhibit qualitatively the same ME spectrum as bulk
iron [26]. The isomer shift of the ME spectrum depends on the
chemical state (chemical shift) and on the atomic vibrations of
the ME atoms (second order Doppler shift and zero point motion
shift). The isomer shift of these clusters merely differ from
that of the bulk value. The shift of the large and the small
clusters amounts to +0.14 mm/s with respect to α-iron standard
at 300 K [31]. However, the expected isomer shift, calculated by
means of the second order Doppler effect on the assumption that
the lattice vibrations of the metal are strongly influenced by
the atomic vibrations of the zeolite matrix, amounts to

+0.16 mm/s related to α-iron. This agreement between calculated and measured isomer shift values confirms the model of phonon interaction between cluster and support.

Due to this matrix effect, especially the vibrations of those atoms which are located in the interface between the nucleus of the cluster and the wall of the cavities should give rise to an asymmetry of the atomic vibrations. This should result in an asymmetry of the γ-resonance absorption especially for the fraction of the very small clusters which contain mostly surface atoms. In fact, it was shown recently that small particles of iron supported on A-zeolites (mean particle size: $v^{1/3} \sim 1.0$ nm) give rise to an asymmetric doublet [31].

Electronic and Magnetic Properties

Because of the existence of a correlation between the electron configuration of a ME atom and the isomer shift of a ME spectrum, it should be interesting to investigate whether or not the isomer shift of the ME spectrum of metal aggregates on zeolites corresponds to that of bulk iron.

As recent measurements show, the isomer shift of small iron clusters agrees well with the value of bulk iron, if corrections for the lattice coupling have been made (see above). However, this does not necessarily mean that the electron configurations are identical. Namely, to conclude from the isomer shift to the electron configuration, the knowledge of the number of electrons per iron atom is necessary. Therefore, the basic question arises, whether there is an electron transfer from the metal particle to the support. This problem is also important in respect to the catalytic properties. Here, the discussion is still controversal [11].

Thus, the next question to be considered is the nature of the bonding between the metal crystallite and the nonmetallic support. By using alkaline vapour reduction [24], the generation of an oxidized metal-support interface will presumably be prevented. The problem, however, we are concerned with, refers to the possibility of partial electron transfer between the metal and electron acceptor sites of the support to such an extent, that the gross electronic properties of the metal particle may be modified. For instance, the minimum electron affinity of the support necessary for Ag_4^+ to be formed from Ag_4 amounts to 4 - 8 eV [32].

On the other hand, the ionization potentials for the aromatic hydrocarbons, which successfully have been used as adsorbents to identify surface electron acceptor sites in insulating oxides, usually lie in the region of 6 to 8 eV. Thus, it is quite reasonable to assume electron transfer to some extent. Accordingly, in order to analyze the ME spectrum, we need additional information whether or not electron transfer has taken place. This may be obtained with the help of measurements of the spontaneous magnetization. In this case, new problems arise: According to ab initio calculations, the absolute value of the spontaneous magnetization at 0 K depends on the particle size [33]. In addition, for higher temperatures the temperature function of the spontaneous magnetization is reported to depend strongly on the particle size and on the particle shape [34]. Until today, however these results could not be proven by experiments.

The spontaneous magnetization and the volume of small particles of a ferromagnetic supported metal can be calculated from static measurements of the magnetization [35-38].

For small, non-interacting ferromagnetic particles with negligible small magnetic anisotropy the magnetic moment of n_i particles m_i would be

$$m_i = M_{sp} \cdot n_i \cdot v_i \left[\coth \left(\frac{M_{sp} \cdot v_i \cdot H}{kT} \right) - \frac{kT}{M_{sp} \cdot v_i \cdot H} \right] \tag{1}$$

where M_{sp} is the spontaneous magnetization per unit volume, and H the magnetic field strength. Both low and high H/T approximations can be made.

At low values of H/T the equation (1) reduces to

$$m_i = \frac{M_{sp}^2 \cdot n_i \cdot v_i^2 \cdot H}{3 \ kT} \tag{2}$$

For a collection of particles of various sizes, the total magnetic moment $m = \sum_i m_i$, so that

$$m = \frac{M_{sp}^2 \cdot H}{3kT} \sum_i n_i \cdot v_i^2 \tag{3}$$

However, the saturation moment m_s is given by

$$m_s = M_{sp} \sum_i n_i \cdot v_i \tag{4}$$

Thus, by equating m/m_s to σ/σ_s or to M/M_s

$$\frac{M}{M_s} = \frac{\sigma}{\sigma_s} = \frac{M_{sp}}{3k} \left(\frac{H}{T} \right) \left(\frac{\sum_i n_i \cdot v_i^2}{\sum_i n_i \cdot v_i} \right) \tag{5}$$

On the other hand, at high values of H/T

$$\frac{M}{M_s} = \frac{\sigma}{\sigma_s} = 1 - \frac{k}{M_{sp}} \left(\frac{T}{H}\right) \left(\frac{\sum_i \cdot n_i \cdot v_i}{\sum_i \cdot n_i}\right) - 1 \tag{6}$$

where M is the magnetization per unit volume, σ the magnetization per unit mass, and M_s and σ_s are the corresponding saturation values.

Usually, there is a distribution of particle sizes in zeolites [24,26]. A large amount of clusters is encapsulated inside the cavities of the matrix, whereas a small amount is located in enlarged pores (which are produced by local distortion of the secondary structure)(Fig. 3) or on the external surface of the zeolite grain.

At low values of H/T the magnetization is mostly affected by the largest particles whereas at high H/T values the magnetization is affected by the smallest particles. In case of this kind of distribution, the magnetization curve cannot be fitted by the simple function equation (1). Therefore, we usually approximated the measured magnetization isotherm by using a set of equations such as (1) with various values of v_i. For the low temperature isotherm we obtain the particle size distribution of the small clusters, whereas for the high temperature isotherm we get the size distribution of the larger particles; both with high accuracy. Table 1 shows the mean particle size of iron clusters encapsulated in A-zeolite as was calculated from the size distribution of the magnetization isotherm at 9.8 K between $v^{1/3}$ = 0.56 nm and $v^{1/3}$ = 1.63 nm. $v^{1/3}$ = 0.56 nm corresponds to 15 atoms and $v^{1/3}$ = 1.63 nm corresponds to 369 atoms, respectively (bcc Fe with lattice constant 0.2866 nm). To a first approximation, the volume of each fraction of the total size distribution has been calculated from the magnetic moments $m_i = M_{sp}v_i$ on the assumption that M_{sp} is equal to that of bulk iron. This leads to an extrapolated value of the saturation magnetization, corresponding to that calculated due to the iron content in the sample.

This is in agreement with the result of a MES investigation where the degree of reduction has been found to be 100% [39,40]. To obtain further proof whether the choice of the bulk value of M_{sp} is correct or not, an independent method for determining the size distribution is necessary. Adolph performed magnetic neutron small angle scattering (nsas) experiments at 35 K on equally treated samples of Fe(0)-A-zeolites. The mean particle size established from the size distribution between $v^{1/3}$ = 0.56 nm and $v^{1/3}$ = 1.63 nm is also listed in Table 1.

TABLE 1

Radius $\bar{R}$ of the mean volume for the fraction of iron particles in A-zeolites with volumes between v = 0.56^3 nm^3 and v = 1.63^3 nm^3.- A comparison between magnetic and neutron small angle scattering measurements ('nsas').

Sample No.	measurement technique	$\bar{R}$ (nm)
1	nsas	0.60 ± 0.02
2	nsas	0.59 ± 0.02
1	magnetic	0.59 ± 0.02

In addition to particle sizes ranging from $v^{1/3}$ = 0.56 nm to $v^{1/3}$= 1.63 nm, smaller and larger particles have been found. The magnetization isotherm of the smaller clusters is linear at 9.8 K. According to the 100% degree of reduction, found by MES, this magnetization cannot be due to iron ions, but must be due to metal aggregates. For the magnetization curve of very small clusters containing a few atoms, the approximation J → ∞ is no longer valid J is the total angular momentum. Instead of the classical Langevin function - equation (1) - the Brillouin function has to be applied. The measured magnetization of the very small clusters at 9.8 K corresponds to a Brillouin function with J = 2. This value is in agreement with a ferromagnetic exchange coupling between four iron atoms. Thus, 42 wt% of the iron, seem to be isolated in the cavities of A-zeolite in the state of a four atom aggregate, probably with tetrahedral geometry [41]. These very small particles are not detectable by 'nsas' [40].

In addition to the small and the very small 'microcrystals' approx. 7% of the iron particles are larger than v = 1.6^3 nm^3 and smaller than v = 9.3^3 nm^3 as was proven by the magnetic and 'nsas' measurements.

The good agreement between the two independent methods indicate that in fact the absolute value of the spontaneous magnetization is the same as in bulk iron. It is important to know that this result was obtained by 'in situ' measurements.

At higher temperatures, the mean magnetic moment of the iron particles $m_i = M_{sp}v_i$ depends on the temperature [40]. This may either be due to the fact that the high-argument- and low-argument-approximation of the Langevin function yields a differently defined mean value. Alternatively it may be due to a different temperature function of the spontaneous magnetization of the large amount of very small clusters containing only four atoms.

It has been reported that 0.6^3 nm^3 Co particles encapsulated in the pore system of mordenite show the absolute value and the temperature function of the bulk metal [25,29]. However, the absolute value of M_s of Co particles encapsulated in Y-zeolites is much less than the expected value [42]. In both cases, the measurements were not performed in situ. Therefore, the observed deviation from bulk-like behaviour may be due to some traces of oxygen which could have entered during the filling of the sample holder used for the magnetic measurements. Oxygen diffusion to the metal surface is easy in case of the three dimensional channel system in Y-zeolites whereas in case of mordenite one particle on each side of the channel may protect the particles behind from instant surface oxidation.

From the fact, that the spontaneous magnetization of even 0.6^3 nm^3 Co in mordenite and Fe in A-zeolite does not differ from massive metal we can conclude that no electron transfer from the metal to the support has taken place.

This is a surprising result because the above mentioned rough estimations predicted such effects. Furthermore, even for clusters of noble metals supported on zeolites such as Pt the clusters are reported to carry a positive charge [43].

In case of Pd clusters encaged in zeolites, however, no support effect on the electronic properties could be detected [43].

Recent ESCA investigations by Lunsford et al. [11] could not establish the results of Vedrine et al. [43] concerning the Pt particles.

Hence, up to today there is no evidence for electron transfer between noble metals and zeolite matrices. Our results on Co supported on mordenite and Fe encapsulated in A-zeolite have shown, that even non-noble metals do not exhibit such charge transfer. It has to be mentioned that the Lewis- as well as Brönstedt-acidity of A-zeolite and mordenite is quite different.

From the fact, that no charge transfer from Fe clusters to the matrix of A-zeolite has taken place it is clear, that the number of electrons per iron atom remains the same as in bulk iron. Therefore, from the value of the isomer shift of the ME spectrum of these clusters we must conclude, that the electron configuration is the same (within the experimental error) as in the massive metal.

The reported results on particle support interaction flash a new light on the old question whether the unusual catalytic behaviour of metal aggregates supported on zeolites are due to size effects or electron transfer. Now it seems to be very likely that the size effect plays the major role.

Especially the probability of the stabilization of microclusters containing four [40] or six [7] uncharged and three [44] or fourteen [45] charged atoms, respectively, underlines the necessity for developing atomistic models based on the real metal-zeolite system. For this purpose, more structural information concerning the location of the metal atoms with respect to each other as well as to the matrix are necessary.

Finally, theoretical and experimental work has to be done on microcluster-product-reactand complexes for the understanding of a particular catalytic reaction. There is obviously still a long way to reach this goal.

CONCLUSION

The first aspect of this work was to demonstrate how many atoms are necessary to form a metal aggregate with bulk-like properties. It has been found that the electronic and magnetic properties of clusters containing more than approx. fifteen atoms seem to be necessary to obtain this state. The second aspect can be summarized as follows: it could be shown, that electron microscopic, magnetic and Mößbauer spectroscopic methods are very reliable for the characterization of metal aggregates larger than approx. 0.6^3 nm^3. The last aspect concerning the 'in situ' measurements still has to be proven for the transition-metal-zeolite system. However, in principle magnetic and MES measurements can even be performed while the catalytic reaction is taking place.

REFERENCES

1 J.R. Anderson, Structure of Metallic Catalysts, Academic Press, London, 1975.
2 A.P. Bolton, in J.A. Rabo (Ed.), 'Zeolite Chemistry and Catalysis', ACS Monograph 171 (1976) 714.
3 C. Naccache, in L.V. Rees (Ed.), Proc. of the Vth Int. Conf. on Zeolites, Heyden, London, 1980, p. 592.
4 Kh.M. Minachev, Ya.I. Isakov, in J.A. Rabo (Ed.) 'Zeolite Chemistry and Catalysis', Chapter 10, ACS Monograph 171 (1976)552.
5 R. van Hardeveld, F. Hartog, Surface Sci. 15 (1969) 189.
6 J.A. Rabo, V. Schomaker, P.E. Pickert, in Proc. 3rd Int. Congr. Catalysis, Vol. 2, North Holland, Amsterdam 1965, p. 1264.
7 R.A. Dalla Betta, M. Boudart, in Proc. 5th Intern. Congr. Catalysis, Vol. 2, North Holland, Amsterdam, 1973, p. 1329.
8 J.B. Uytterhoeven, Acta Phys. et Chem. Szeged 24 (1978) 53.
9 P. Gallezot, Catal. Rev. Sci. Eng. 20 (1979) 121.
10 P. Gallezot, in B. Imelik et al. (Eds.), Catalysis by Zeolites, Elsevier, Amsterdam 1980, p. 227.
11 J.H. Lunsford, D.S. Treybig, J. Catal. 68 (1981) 192-196.
12 D. Delafosse, in B. Imelik et al. (Eds.), Catalysis by Zeolites Elsevier, Amsterdam 1980, p. 235.
13 P.N. Galich, V.S. Gutyrya, A.A. Galinski, in L.V. Rees (Ed.), Proc. of the Vth Int. Conf. on Zeolites, Heyden, London, 1980, p. 661.
14 Ch. Minchev, V. Kanazirev, L. Kosova, V. Penchev, W. Gunsser, F. Schmidt, in L.V. Rees (Ed.), Proc. of the Vth Int. Conf. on Zeolites, Heyden, London, 1980, p. 355.
15 Kh.M. Minachev, Ya. I. Isakov, V.P Kalinin, in L.V. Rees (Ed.), Proc. of the Vth Int. Conf. on Zeolites, Heyden, London, 1980, p. 866.
16 D. Fargues, F. Vergand, E. Belin, C. Bonnelle, D. Olivier, L. Benneviot, M. Che, Surface Sci. 106 (1981) 239.
17 F. Schmidt, T. Meeder, Surface Sci. 106 (1981) 397.
18 N. Jaeger, U. Melville, R. Nowak, H. Schrübbers, G. Schulz-Ekloff, in B. Imelik et al. (Eds.), Catalysis by Zeolites, Elsevier, Amsterdam 1980, p. 335.
19 P.A. Jacobs, H. Nijs, J. Verdonck, E.G. Derouane, J.P. Gilson, A.J. Simoens, in JCS Faraday I 75 (1979) 1196.
20 M. Grunze, D. Wright, G. Schulz-Ekloff, Zeolites 2 (1982) in press.
21 D. Exner, N. Jaeger, R. Nowak, H. Schrübbers, G. Schulz-Ekloff, J. Catal. 74 (1982) in press.
22 R.P. Messmer, Surface Sci. 106 (1981) 225.
23 F. Schmidt, W. Gunsser, A. Knappwost, Ber. Bunsenges. Phys. Chem. 77 (1973) 1022.
24 F. Schmidt, W. Gunsser, A. Knappwost, Z. Naturforsch. 30a (1975) 1627.
25 F. Schmidt, EUCHEM Conference on Dispersed and Colloidal Metal Particles in Catalysis, Namur 1977, unpublished.
26 F. Schmidt, W. Gunsser, J. Adolph, ACS Symposium Series 40 (1977) 291.
27 K.M. Sancier, W.E. Isakson, H. Wise, in E.L. Kugler and F.W. Seffgen, Hydrocarbon Synthesis from Carbon Monoxide and Hydrogen, Advances in Chemistry Series 178 (1979) 129.
28 F. Schmidt, W. Gunsser, W. Vollberg, J. Magn.Magn.Mater. 6 (1977) 220.
29 W. Vollberg, Thesis, University of Hamburg, 1977.
30 J.W. van Wieringen, in Physics Lett. 26A (1968) 370.
31 F. Schmidt, J. Adolph, Journal de Physique, in press.

32 J.R. Anderson, Structure of Metallic Catalysts, Academic Press, London, 1975, p. 278.

33 D.R. Salahub, R.P. Messmer, Surface Sci. 106 (1981) 415.

34 K. Binder, Z. angew. Phys. 30 (1970) 51.

35 P.W. Selwood, Adsorption and Collective Paramagnetism, Academic Press, New York, 1962.

36 C.P. Bean, I.S. Jacobs, J. Appl. Phys. 27 (1956) 1448.

37 A. Knappwost, Z. Elektrochem. 61 (1957) 1328.

38 A. Knappwost, Z. Elektrochem. 63 (1959) 278.

39 F. Schmidt, J. Adolph, J. Magn. Magn. Mater. 15-18 (1980) 1115

40 J. Adolph, Thesis, University of Hamburg, 1982.

41 H.H. Dunken, Z. phys. Chem. Leipzig 246 5/6 (1971) 329-341.

42 F. Schmidt, W. Vollberg, to be published.

43 J.C. Vedrine, M. Dufaux, C. Naccache, B. Imelik, JCS Faraday I 74 (1978) 440.

44 L.R. Gellens, W.J. Mortier,* R.A. Schoonheydt, J.B. Uytterhoeven, J. Phys. Chem. 85 (1981) 2783-2788.

45 Y. Kim, K. Seff, J. Am. Chem. Soc. 100 (1978) 175.

ELECTRON MICROSCOPICAL ANALYSIS OF MONODISPERSED Ni AND Pd IN A FAUJASITE X MATRIX

D. Exner[1], N.I. Jaeger[1], K. Möller[1], R. Nowak[1], H. Schrübbers[1], G. Schulz-Ekloff[1] and P. Ryder[2]

[1]Forschungsgruppe Angewandte Katalyse, Fachbereich 3 - Chemie, Universität Bremen

[2]Forschungsgruppe Licht- und Elektronenmikroskopie, Fachbereich 2 - Physik, Universität Bremen, D-2800 Bremen 33

ABSTRACT

Monodispersed metal phases with crystallite sizes exceeding supercage dimensions by far can be contained and stabilized within the faujasite X matrix. This is demonstrated in the case of Pd by transmission electron microscopy in combination with electron and X-ray diffraction. From the similarity of the monodispersion and distribution for Ni, Pt and Pd a common growth mechanism for the single crystal metal particles on the basis of a secondary reduction process is suggested.

INTRODUCTION

The zeolite matrix proves to be especially well suited for the preparation of highly dispersed phases of catalytically active metals. In most cases metal agglomerates are found within supercage dimensions as long as they are located inside the zeolite framework (ref. 1-5). However, crystallite sizes exceeding supercage dimensions by far can also be accomodated within the faujasite framework as has been demonstrated recently for a number of metals, namely Pt (ref.6), Ru (ref.7,8), Pd (ref.4) and Ni (ref. 9,10). The preparation of monodispersed metal phases with narrow particle size distribution and variable size range within the faujasite structure offers the opportunity to study particle size effects and support interaction on stability, activity and selectivity of the catalyst in more detail (ref.4, 11-14).

However, partial destruction of the host lattice has to be considered and raises a number of questions, e.g. regarding the growth mechanism of the metal particles within the matrix, the stabilization of the system during catalytic reactions and the possible alteration of the properties of the support, which is itself catalytically active. In this context the results of transmission electron microscopy (TEM), electron diffraction and X-ray diffraction used to characterize monodispersed phases of Ni and Pd with crystallite sizes between 6 and 10 nm and located within the faujasite X matrix will be presented and discussed.

EXPERIMENTAL

Samples

The NaX zeolite $Na_{86}/(AlO_2)_{86}(SiO_2)_{106}/\cdot nH_2O$ was prepared by hydrothermal crystallization (ref. 15,16). Details of the preparation and characterization of monodispersed nickel in reduced NiCaX with particle sizes in the range of 10 nm within the faujasite matrix have been given elsewhere (ref. 10,11). Samples with 4.2 wt% were used. Palladium loading was achieved via ion exchange from 0.05 mole$\cdot dm^{-3}$ solutions of the tetrammine chloride at room temperature. The partially dehydrated (12 h, room temperature) samples (2g) have been heated ($5K\cdot min.^{-1}$) in streaming Ar (50 ml$\cdot$ min.$^{-1}$) up to $300^{o}C$, which led to the formation of the monodispersed metal phase. The procedure was carried out in a fluidized bed reactor. The determination of the metal dispersity via oxygen chemisorption required additional and repeated treatments with hydrogen (ref. 17). Samples with 15 wt% Pd were used.

Methods

The metal particle size distributions have been determined by TEM. The sample preparation has been described elsewhere (ref. 6, 10). Bright and dark field micrographs and selected area electron diffractions were obtained with a Zeiss EM 10.

The crystallinity of the host lattices was controlled by X-ray powder diffraction and by nitrogen physisorption capacities (ref. 17).

RESULTS

Typical transmission electron micrographs of monodispersed metal phases, i.e. metal particles exclusively inside the zeolite matrix and with a narrow particle size distribution around 10 nm, are given in Fig. 1 for nickel and in Fig. 2 for palladium.

In Fig. 1 phase contrast imaging of the zeolite lattice enca-
ging the metal particles confirms that the zeolite structure has
been largely maintained. The metal aggregates can be clearly distinguished from the support. Fig. 2 demonstrates the apparent random distribution of the metal particles across the micrograph,
which is typical for the observed metal phases.

Selected-area electron diffraction patterns taken from the Pd-
loaded samples with different aperture sizes (see Fig. 3) showed
a number of sharp rings superimposed on a diffuse background. The
rings can be indexed in agreement with the face-centred cubic crystal structure of metallic palladium. The lattice parameter also
agreed with the literature value for Pd (0.389 nm) within the
limits of the experimental error of these measurements (± 0.005 nm).

In the diffraction patterns taken with the large aperture (see
Fig. 3a) the diffracted intensity is uniformly distributed around
the rings. In the case of the smaller aperture (see Fig. 3b),
where the individual diffraction spots are distinguishable, there
is also no indication of a preferred orientation.

The dark-field micrograph shown in Fig. 4 was taken with the
(111) Bragg reflection. Only those Pd particles, which are orien-
tated such that a (111) diffracted beam falls within the diffraction aperture appear bright in this micrograph. The fact that the
spatial distribution of these particles is fairly uniform is again
an indication that the Pd crystals have a random orientation,
which would mean that there is no special orientation relationship
between the Pd and zeolite crystal lattices. No diffraction spots
were obtained from the zeolite crystal in this specimen, so that
it was not possible to check the orientation relationship directly.

Apart from the distortions due to lens aberrations, the bright
spots in the dark-field micrograph of Fig. 4 have the same size as
the Pd particles seen in Fig. 2. This means that the orientation
in each particle is uniform throughout, i.e. that each particle is
a single crystal and that a grape structure can be ruled out. On
account of the high particle density it was not possible to obtain
a diffraction pattern from a single particle.

The presence of crystalline Pd was also confirmed by X-ray diffraction, as shown in Fig. 5. In the X-ray diagrams of the Pd-loa-
ded faujasite X, the zeolite lines are found to largely disappear

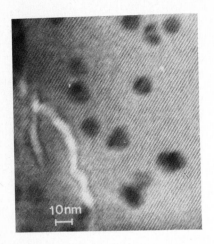

Fig. 1. Transmission electron micrograph of reduced NiCaX 480.000 : 1

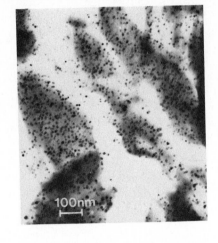

Fig. 2. Transmission electron micrograph of reduced PdX 64.000 : 1

a)

b)

Fig. 3. Electron diffraction pattern of specimen shown in Fig. 2. Diameter of selected area: a) 500 nm   b) 50 nm

Fig. 4. Dark-field image of the same area as shown in Fig. 2, using the (111) Bragg reflection. This image is distorted due to lens aberrations. 64.000 : 1

following the autoreduction procedure (see Fig. 5a) and to reap-
pear following additional treatment under hydrogen. In the Ni-loa-
ded specimens, however, no diffraction lines from metallic Ni were
observed either by electron or X-ray diffraction.

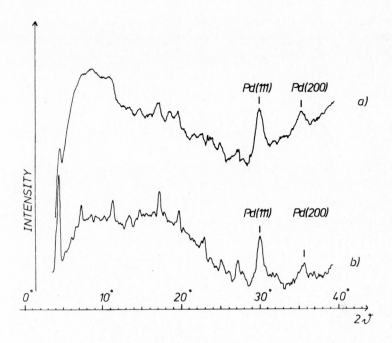

Fig. 5. X-ray diffraction diagrams of PdX following autoreduction
(a) and subsequent hydrogen treatment (b).

DISCUSSION

The experimental results may be interpreted as follows:
(i) The faujasite largely maintains its crystal structure during
metal loading. This has also been confirmed by measuring the nitrogen physisorption capacity of the metal loaded faujasite
samples (ref. 17). However, the fact that in many cases (especially for Pd) X-ray diffraction shows relatively weak, broad zeolite
lines on a high diffuse background level after reduction means
that the zeolite matrix suffers a certain amount of distortion.
The high background level may be attributed in part to incoherent
scattering by very small metal clusters. This effect appears to
be more pronounced for Pd than for Ni, since the former has a

higher X-ray scattering factor. The increase in the relative intensities of the zeolite lines observed following tempering under hydrogen may be attributed to a sintering process, which decreases the number of very small metal aggregates of supercage dimensions and thus the number of lattice distorting and X-ray scattering particles.

(ii) The electron diffraction micrographs obtained from Pd exhibit the face-centred cubic structure of this metal and confirm the results of the X-ray diffraction. In addition, they show that no special orientation relationship exists between the Pd metal particles and the zeolite matrix.

It is to be expected that the Ni particles are also crystalline, although no direct diffraction evidence has been obtained so far. The detection of Ni diffraction lines is more difficult as compared to Pd on account of the lower atomic number.

(iii) The single crystal nature of the Pd particles, revealed from the dark-field imaging, rules out growth mechanisms involving the clustering of independently nucleated smaller particles from contiguous supercages resulting in grape structures, which have been discussed in the case of nickel loaded zeolite Y (ref. 18). Such a structure could also be excluded via a comparative determination of the particle sizes by oxygen chemisorption and transmission electron microscopy (ref. 17).

(iv) The similarity of the monodispersion and distribution of the metal crystallites within the zeolite matrix observed for Ni, Pd and also Pt (ref. 6) suggests a similar growth mechanism. This mechanism might be connected with a secondary reduction process, because the primary reduction processes for Pd and Ni differ extremely.

For the hydrogen reduction of the nickel loaded faujasites the preferential reduction of octahedrally coordinated Ni^{2+} ions in the SI positions has been deduced from experimental results (ref. 10, 11). This could be further assured by the strict correlation between the starting activity of reduced nickel faujasites in the methanation reaction, on one hand, and the accessibility of SI positions for the nickel ions, on the other hand (ref. 19).

The thermally induced primary reduction of the palladium loaded faujasite includes preferentially the following step

$$Pd^{2+}\text{-}(\dot{N}H_3)_x \rightarrow Pd^0\text{-}(NH)_y + 2H^+ + zNH_3 + \frac{(x-y-z)}{2}N_2 + \frac{3(x-z)-y-2}{2}H_2$$

This autoreduction mechanism could be elucidated by thermal analysis coupled with mass spectrometry (ref. 20).

The secondary reduction mechanism is based on the hydrogen activating properties of the metal aggregates formed by the primary process and is controlled by the diffusion of metal ions towards these aggregates. The secondary reduction process creates a high concentration of protons in the neighborhood of the growing metal particles and accounts for the partial destruction of the zeolite lattice in order to accomodate the crystallite. This proposal follows suggestions forwarded by Beyer et al.(ref. 21) and Verdonck (cf. ref. 3).

(v) Monodispersed metal phases with crystallite sizes exceeding supercage dimensions by far can be contained and stabilized within the faujasite X matrix. This has also been confirmed by XPS-measurements, yielding metal signal attenuations for the samples discussed here (ref. 22).

ACKNOWLEDGEMENT

The authors wish to thank Prof. Dr. H. Lechert for his critical revision of the paper and Mr. G. Ernst for carrying out the electron diffraction measurements.

REFERENCES

1   D.W. Breck, Zeolite Molecular Sieves - Structure, Chemistry and Use, New York, 1974, p. 519
2   Kh. Minachev and Ya.I. Isakov, Zeolite Chemistry and Catalysis (ed. J.A. Rabo) A.C.S. Monograph 171 (1976), 552
3   J.B. Uytterhoeven, Acta Phys. et Chem., Szeged 24 (1978) 53
4   P. Gallezot, Stud. Surf. Sci. Catal. (Zeolite Catal.) 5 (1980), 227
5   D. Delafosse, Stud. Surf. Sci. Catal. (Zeolite Catal.) 5 (1980), 235
6   D. Exner, N. Jaeger and G. Schulz-Ekloff, Chem. Ing. Techn. 52 (1980), 734
7   J.J. Verdonck, P.A. Jacobs, M. Genet and G. Poncelet, J.C.S. Faraday I 76 (1980), 403
8   H.H. Nijs, P.A. Jacobs, J.J. Verdonck and J.B. Uytterhoeven, Stud. Surf. Sci. Catal. (Metal Clusters) 4 (1980), 479
9   P.A. Jacobs, H.H. Nijs, J. Verdonck, E.G. Derouane, J.P. Gilson and A.J. Simoens, J.C.S. Faraday I 75 (1979), 1196
10   D. Exner, N. Jaeger, R. Nowak, H. Schrübbers and G. Schulz-Ekloff, J. Catal. 74 (1982), 188
11   N. Jaeger, U. Melville, R. Nowak, H. Schrübbers and G. Schulz-Ekloff, Stud. Surf. Sci. Catal. (Zeolite Catal.) 5 (1980), 335
12   J.A. Rabo, V. Schomaker and P.E. Pickert, Proc. 3rd Int. Cong. Catalysis, Vol. 2, North Holland, Amsterdam 1965, p. 1264

13 Tran Manh Tri, J. Massardier, P. Gallezot and B. Imelik, Stud. Surf. Sci. Catal. (Zeolite Catal.) 5 (1980), 279
14 P.A. Jacobs, Stud. Surf. Sci. Catal. (Zeolite Catal.) 5 (1980), 293
15 D.W. Breck and E.M. Flanigan, Molecular Sieves, Society of Chemical Industry, London 1968, p. 47
16 H. Kacirek and H. Lechert, J. Phys. Chem. 79 (1975), 1589
17 S. Briese-Gülban, H. Kompa, H. Schrübbers and G. Schulz-Ekloff, React. Kinet. Catal. Lett.   in press
18 N.P. Davidova, M.L. Valcheva and D.M. Shopov, Stud. Surf. Sci. Catal. (Zeolite Catal.) 5 (1980), 285
19 H. Schrübbers, Thesis, Fachbereich 3 (Chemie), Universität Bremen, 1982
20 D. Exner, N. Jaeger, K. Möller and G. Schulz-Ekloff, to be published
21 H. Beyer, P.A. Jacobs and J.B. Uytterhoeven, J.C.S. Faraday I 72 (1976), 674
22 G. Schulz-Ekloff, D. Wright and M. Grunze, Zeolites 2 (1982) in press

INVESTIGATIONS OF THE AGGREGATION STATE OF METALS IN ZEOLITES BY MAGNETIC METHODS

W. ROMANOWSKI

Institute of Low Temperature and Structure Research, Polish Academy of Sciences, 50950 Wrocław, P.O. Box 937,(Poland)

ABSTRACT

It has been shown that the application of the static magnetic measurements can give much information about the dispersion, reduction degree, and aggregation processes occuring in metal-zeolite systems containing a ferromagnetic metal. In the case of nickel dispersed in A, X and Y zeolites, at least two populations of nickel particle sizes could be distinguished by the analysis of magnetization curves: larger particles on the external surface of zeolite grains, and small ones, about 1 nm in diameter in the large cavities of zeolite lattice.

INTRODUCTION

Metal/zeolite systems are bifunctional catalysts capable of acid-base as well as metallic functions. During last ten or so years we investigated both these activities of the system nickel/zeolite A, X and Y synthetic zeolites, and especially the possible correlation of dispersion and reduction extent of the metal with these activities.

MAGNETIC PROPERTIES OF NICKEL/ZEOLITE SYSTEMS

We used static magnetic method to determine both the reduction degree after a definite treatment of our samples with gaseous hydrogen at elevated temperature, and the nickel particle size. The former of these quantities can be easily determined basing on the saturation magnetization value, while the latter requires some more elaborate analysis of the magnetization curve of the system containing very small ferromagnetic particles dispersed in a diamagnetic matrix. The properties of such a system are called superparamagnetic.

The magnetization of a superparamagnetic system can be described by the reduced Langevin equation

$$\frac{\sigma}{\sigma_\infty} = cth \frac{\mu_f}{k} \cdot \frac{H}{T} - \frac{kT}{\mu H} = L\left(\frac{\mu_f \cdot H}{kT}\right) \tag{1}$$

where σ is the magnetization at the magnetic field strength H, and temperature T, σ_∞ - saturation magnetization, μ_f - magnetic moment of a small ferromagnetic particle equal to the sum of atomic magnetic moments, k - Boltzmann constant.

Basing on the equality $\mu_f = v \cdot I$ (v - the volume of a particle, I - saturation magnetization per unit volume) one can calculate v and thus the size (diameter) of the particles. In real systems, however, small ferromagnetic particles have, in general, different sizes. Additionally, sometimes only a part of them have sufficiently small "superparamagnetic" sizes, conforming to the Langevin equation (1). Thus, generally, the magnetization curve within the range of higher fields and below the ferromagnetic Curie point will be expressed by following equation

$$\frac{\sigma}{\sigma_{\infty}} = n_o \left(\frac{\sigma}{\sigma_{\infty}}\right)_o + n_1 \int_{v_1}^{v_2} L \left(\frac{v \cdot I_s \cdot H}{kT}\right) \; F(v) \; dv \qquad , \tag{2}$$

where n_o is the fraction of normal ferromagnetic particles, n_1 - the superparamagnetic fraction, and $F(v)\,dv$ - volume distribution of superparamagnetic particles.

It has been shown previously that nickel/zeolite systems tend to contain almost always a significant part of "normal ferromagnetic" nickel particles i.e. particles large enough to saturate at not too high fields (ref.1,2,3).

When the size distribution function is polymodal in character and the range of sizes around each mode is sufficiently narrow, the equation (2) can be reasonably approximated by

$$\frac{\sigma}{\sigma_{\infty}} = n_o \left(\frac{\sigma}{\sigma_{\infty}}\right)_o + n_1 \; L \left(\frac{\mu_{f_1} H}{kT}\right) + n_2 \; L \left(\frac{\mu_{f_2} H}{kT}\right) + \ldots \tag{3}$$

where the second and following terms express the contribution of particles of mean magnetic moment μ_{f_1}, μ_{f_2} and so on. The number of the representative magnetic moments, by which an experimentally found magnetization curve of Ni/zeolite samples can be described, is mostly not greater than two.

The value of saturation magnetization σ_{∞} is measured after prolonged heating of the reduced nickel/zeolite samples at the temperature higher than that of reduction (723 K) in vacuo; otherwise it is approached at very high fields, and the extrapolations from the data obtained at the fields of the order of 1 T are uncertain.

Magnetization curves of a series of Ni/zeolite catalysts, obtained by the hydrogen reduction at various temperatures have been measured at temperatures of 77 and 292 K and in the fields up to 0.7 T in order to obtain magnetization values within the broad range of H/T (ref.4). All these curves could be fitted to the equations of the form of equation (3) indicating the presence of some part

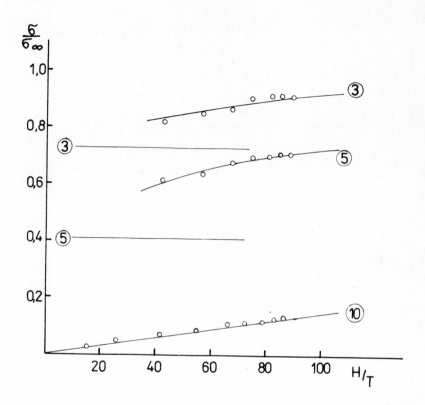

Fig. 1. Reduced magnetization curves of samples 3, 5 and 10 (Table 1).
Continuous lines - equations of the form (3), circles - experimental values.
Horizontal lines indicate "massive ferromagnetic" fraction of Ni .

of "massive ferromagnetic" particles and one or two representative sizes of superparamagnetic particles (see Table 1 below and Fig. 1).

Representative radiuses of superparamagnetic particles r_1 and r_2 were calculated assuming the moment 0.61 Bohr magnetons per one nickel atom; from the moments listed in the Table 1 the number of atoms in the particles, their volume
and radius could be calculated.

From the data of the Table 1 it is evident that the time of reduction influences the distribution of particle sizes less than the temperature (samples 1-6).
One sees also that the calculation at 693 K of the exchanged and dried prepara-
tions before reduction increases the fraction of smallest superparamagnetic
particles and decreases their mean dimension. This is probably caused by a more
uniform distribution of the exchanged Ni^{2+} - ions in the whole volume of zeolite
grains, brought about by calcination. The comparatively high degree of reduction

TABLE 1
Approximate particle size distribution and reduction degree of Ni/zeolite catalysts

Sample No	Zeolite	Nickel content before reduction weight %	Time and temperature of reduction	"massive ferromagnetic" n_0	n_1	μ_{f_1} Bohr magnetons	n_2	μ_{f_2} Bohr magnetons	r_1 nm	r_2 nm	Reduction degree of nickel %
1	A	7.81	5 hrs 633 K	0.48	0.28	593	0.24	65.8	1.36	0.66	67.5
2			10 hrs 633 K	0.59	0.31	701	0.10	64.7·	1.44	0.65	66
3	X	11.95	5 hrs 633 K	0.73	0.27	485	–	–	1.27	–	20.5
4			10 hrs 633 K	0.74	0.26	723	–	–	1.45	–	32
5	Y	5.93	5 hrs 633 K	0.41	0.29	776	0.30	129.5	1.48	0.82	17
6			10 hrs 633 K	0.50	0.20	723	0.30	151	1.45	0.86	18.5

Samples heated 4 hrs at 693 K in air before reduction:

Sample No	Zeolite	Nickel content before reduction weight %	Time and temperature of reduction	"massive ferromagnetic" n_0	n_1	μ_{f_1} Bohr magnetons	n_2	μ_{f_2} Bohr magnetons	r_1 nm	r_2 nm	Reduction degree of nickel %
7	X	5.59	4 hrs 623 K	0.08	0.92	74.4	–	–	0.68	–	17
8			4 hrs 653 K	0.14	0.86	179	–	–	0.91	–	29
9			4 hrs 653 K+ 4 hrs 693 K	0.63	0.37	396	–	–	1.19	–	29
10	Y	5.97	3 hrs 573 K 3 hrs 573 K+	0	1.00	68	–	–	0.66	–	39
11			4 hrs 623 K+ 4 hrs 653 K+ 4 hrs 693 K	0.64	0.36	180	–	–	0.91	–	57

and high content of larger ferromagnetic particles in zeolite A are supposed to be connected with highest degree of lattice destruction during the reduction (detected independently by X-ray diffraction) which facilitates the diffusion and aggregation of nickel atoms and small clusters formed during the reduction. On the other hand, nickel in Y zeolite, which preserves the crystalline structure during the reduction at temperatures indicated (ref.3), show smallest dimension of superparamagnetic nickel particles and, except at highest reduction temperatures, low content of "normally ferromagnetic" ones.

The fraction of smaller superparamagnetic particles r below 1 nm is believed to be situated in big zeolite cavities, whose diameters are over 1 nm, with possible extensions to the neighbouring cavities of the same kind, connected by windows of 0.7 nm diameter in faujasites. The larger superparamagnetic particles $r = 1.2 - 1.5$ nm and those termed "normal ferromagnetic" can not be accommodated inside the zeolite framework and thus are placed on the external surface of zeolite grains. They are probably formed by diffusion of small particles toward the surface and their coalescence. The diffusion of metals on the alumosilicate substrate is known to require comparatively small activation energy and its rate is large enough at the temperature of reduction. Only at possibly lowest temperature and short time of reduction it is possible to obtain exclusively small superparamagnetic particles residing in the α-cages of Y zeolite (sample 10).

The particles with the diameter larger than about 3 nm and those still larger ones called "normally ferromagnetic", are hard to distinguish basing on the magnetization curves alone, because they approach saturation magnetization at comparatively low fields. Therefore the proper domain of the magnetic granulometry is the particle size range from below 1 to 3 nm.

The above results show the possibility to control the size of ferromagnetic particles in the zeolite matrix by the proper choice of the reduction conditions. It is thus possible to obtain the smallest dimensions of small particles by the reduction with molecular hydrogen. The preparation of nickel/zeolite samples which contain exclusively Ni particles of about 1 nm in diameter has been recently carried out by the reduction with atomic hydrogen at 273 K (ref. 5), using a rather sophisticated and expensive apparatus. It is obvious, however, that at this reduction temperature the diffusion of reduced nickel and its aggregation must be very slow, thus there is no formation of larger particles.

SYSTEM Cu-Ni/ZEOLITE Y (ref.6)

Another example of the study of migration and aggregation processes involving metals in zeolite by static magnetic method is our investigation of the system Cu-Ni/zeolite Y . The samples have been prepared in the same way, as Ni-zeolite by ion exchange, drying and hydrogen reduction at 623 and 653 K. Magnetization measurements were taken immediately after reduction and degassing of samples in vacuo, and then after definite time interwals in the course of annealing at 723 K. The typical results are shown in the Fig. 2 for the few of the samples investigated.

While the magnetization of samples containing nickel only reached their maximum magnetization value after some 100 hours of annealing and this value reamined constant during further heating at the same temperature (curve 1), for samples with increasing copper content after reaching a maximum, the magnetization decreased during a long time, and in the case of samples containing about 50 %at. of Cu the final magnetization was negligible.

This course of magnetization during sintering can be explained considering the rates of the reduction and of diffusion of both constituents to the outer surface of zeolite grains. Were these rates equal, no change of magnetization, except may be some small its increase owing to the formation of larger alloy crystals would be observed during annealing, because the alloy would be formed already during the reduction. If however, one assumes that the rates of reduction

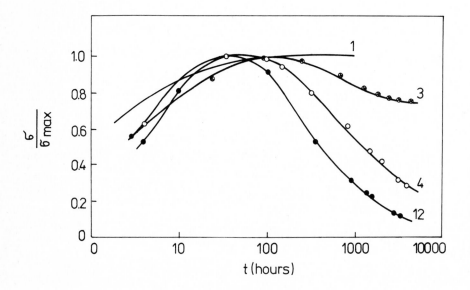

Fig. 2. Magnetization at 292 K of Cu-Ni/zeolite Y samples sintered at 723 K.
1 - pure Ni/zeolite, 3 - 20 at % of Cu, 4 - 44.2 at % Cu, 12 - 54.1 at % Cu.

and diffusion of zero-valent copper is faster than that of nickel, then there will be more copper than nickel on the surface in first stages of reduction, and nickel will be present in the form of separate, very highly dispersed phase (low magnetization) immediately after reduction. During first stages of annealing the nickel will migrate to the surface and its crystals will grow giving rise to the increase of magnetization. This growth can proceed by the surface diffusion of nickel on zeolite and nickel on nickel. After reaching some maximum of magnetization corresponding to the almost separate, comparatively large crystals of copper and nickel, the volume interdiffusion of nickel and copper comes into play during the long time of annealing, leading to the fall of magnetization, the final value of which is that of corresponding homogeneous solid solution. Thus it seems that a difference in the rates of surface diffusion of both species causes the initial rise of magnetization during annealing while its later decrease in a long time is due to a low rate of volume interdiffusion.

The above described experimental results show clearly that metal/zeolite systems working as catalysts at temperatures of the order of few hundreds degrees can undergo profound changes in their aggregation state and composition. The final state, corresponding to thermodynamic equilibrium in these systems may never be reached in the normal span of catalysts life.

REFERENCES

1 W. Romanowski, Z. anorg. allg. Chem. 351 (1967) 180.
2 W. Romanowski, Przemysł Chem. 47 (1968) 741.
3 W. Romanowski, Roczniki Chem. 45 (1971) 427.
4 W. Romanowski, Polish J. Chem. 54 (1980) 1515.
5 M. Che, M. Richard, D. Olivier, J. Chem. Soc. Faraday Trans. I, 76 (1980) 1526.
6 W. Romanowski, Roczniki Chem. 51 (1977) 493.

DIELECTRIC PROPERTIES OF X-TYPE ZEOLITES CONTAINING SMALL METALLIC NICKEL PARTICLES

J.C. CARRU[1] and D. DELAFOSSE[2]

[1] C.H.S. - L.A C.N.R.S. 287 - Equipe "Nouveaux Matériaux" - Bât. P3
Université de Lille 1 - 59655 - Villeneuve d'Ascq cédex (France)

[2] E.R. 133 Réactivité de surface et structure, Laboratoire de Chimie des Solides,
Université P. et M. Curie, 4, Place Jussieu - 75230 - Paris (France)

ABSTRACT

The dielectric method was used in the frequency range 1 Hz-100 MHz to characterize X-zeolites containing Ni^{2+} cations. These cations were reduced by two
different procedures : hydrogen molecules and hydrogen atom beams. From the
analysis of the dielectric spectra, it is possible to obtain qualitative information about the repartition of the metallic aggregates formed in the framework and on aggregate-support and cation-support interactions.

INTRODUCTION

Numerous studies on the properties of small metallic particles included in
various supports have been published up to now. The characterization of these
particles has been developed from magnetic methods (static magnetization,
ferromagnetic resonance) and by conventional techniques (chemisorportion, electron microscopy). The dielectric method developped in the laboratory [1] has
allowed us to obtain information on zeolites containing alkali or alkaline earth
cations at different exchange levels. But, these zeolites contained no metallic
particles. So, it seemed interesting to use samples of zeolites containing metallic particles to study the influence of the formation and the repartition of
these particles on the dielectric behaviour of such zeolites.

In this paper, we give the first results (although partial) obtained on X-
zeolites containing Ni^{2+} cations reduced (or not) by 2 different methods :
hydrogen molecules and hydrogen atom beams.

EXPERIMENTAL

Materials

The samples were prepared by conventional exchange of X-type sodium zeolite
of the Linde Division of Union Carbide with 0.1 normal solutions of Ni^{2+} nitrate. The compositions, determined by chemical analysis were as follows :
$Ni_8 Na_{64} H_6 X$, $Ni_{31} Na_{24} X$. The samples will be hereafter referred to respec-

tively as Ni_8X and $Ni_{31}X$.

Treatment of samples

All the samples were pretreated overnight in vacuo at 673 K. After activation, only the Ni_8X samples were reduced :

- either by hydrogen molecules during about 27 hours at 723 K (sample A) and at 583 K (sample B),

- or by hydrogen atom beams during 5 hours at 273 K (sample C).

In each case, the preparation method of the samples has been described previously [2,3].

Experimental method

After reduction, the samples are placed under vacuo and kept in a sealed tube.

For the dielectric study, about 350 mg of the samples are coated with a silicon oil (of SI 710 type supplied by Rhône-Poulenc) according to a process described elsewhere [4]. Any air contamination being avoided, the dielectric measurements are performed in an air tight cell in the frequency range 1 Hz-100 MHz between 273 K and 353 K.

RESULTS

From the measurement of the complex permittivity $\varepsilon^* = \varepsilon' - j\varepsilon''$, at a given temperature, it is possible to obtain 3 curves :

- ε' (real permittivity) as a function of the frequency
- ε'' (absorption) as a function of the frequency
- ε'' as a function of ε' (Cole and Cole diagram).

Non reduced samples

We give in Figure 1 the curves ε'' versus frequency in logarithmic scales for the samples NaX, Ni_8X and $Ni_{31}X$ at 298 K. We can observe, for each sample, the existence of an absorption domain of energy. The characteristics of these domains depend on the quantity of Ni^{2+} present in the samples. We give also in Figure 2 the Cole and Cole diagram for NaX at 298 K.

Reduced samples

We give in Figure 3, the ε'' spectra at 298 K for the Ni_8X reduced samples (A,B,C). We can see that all the spectra show an absorption domain which depend on the reduction conditions of the nickel cations.

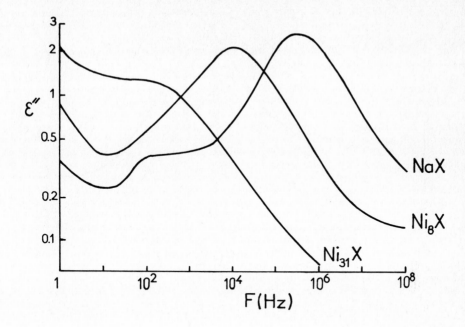

Fig. 1. Dielectric absorption spectra at T = 298 K of NaX, Ni_8X and $Ni_{31}X$.

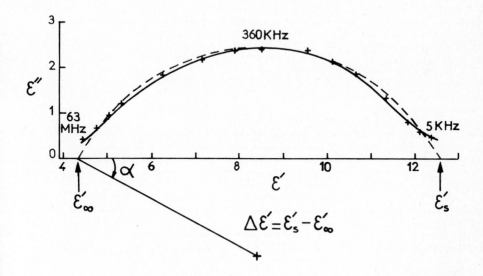

Fig. 2. Cole and Cole diagram of NaX zeolite. T = 298 K.

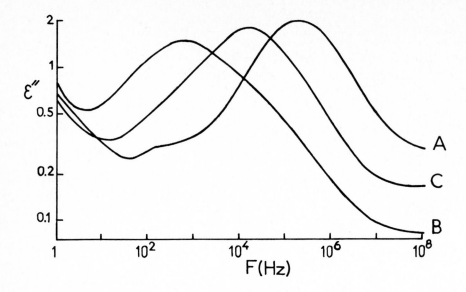

Fig. 3. Dielectric spectra at T = 298 K of Ni_8X reduced samples (A, B and C).

Dielectric parameters for the samples studied

From the dielectric spectra obtained, it is possible to characterize the absorption domain of each sample with 4 parameters :

i) the critical frequency F_c, i.e. the value of the frequency which corresponds to the maximum of the amplitude of the domain (on the ε'' curves).

ii) the activation energy ΔU deduced from the critical frequencies obtained at different temperatures according to $F_c = F_0 e^{-\Delta U/kT}$ [5].

iii) the variation of the real permittivity $\Delta\varepsilon'$ deduced from the Cole and Cole diagram (cf. Fig. 2).

iv) the distribution parameter α also deduced from the Cole and Cole diagram (cf. Fig. 2). α characterizes the spreading in frequency of the domain.

TABLE 1

Values of the dielectric parameters for the zeolites studied

Non reduced zeolites	Ni_8X reduced samples	F_c(Hz)	ΔU (kcal/mole)	$\Delta\varepsilon'$	α (degrees)
NaX		2×10^5	7.4	8.2	28
Ni_8X		13×10^3	8.7	8	35
$Ni_{31}X$		60	11.5	5.7	50
	A	2×10^5	7.8	-	30
	B	700	13.6	7.5	45
	C	17×10^3	9.4	6.6	40

DISCUSSION

Origin of the electrical polarization (non-reduced samples)

It is well admitted by most of the authors [6] that the electrical polarization observed in the frequency range studied, depends on the nature of the cations present in the framework. Thus, for the non-reduced samples (cf. Fig. 1) we have observed that the critical frequency decreased drastically when exchanging the Na^+ cations by Ni^{2+} (by about a factor 3300 between NaX and $Ni_{31}X$). We recall that this critical frequency is linked to the mean frequency of the jumps between the different positions the cations can occupy in the structure [4]. In $Ni_{31}X$, the location of Ni^{2+} cations is as follows : 12 S_I ; 9,5 $S_{I'}$; 6 S_{II} and 4 in the supercages outside the ternary axis [7]. So, the presence of Ni^{2+} with these arrangement modifies strongly the mobility of Na^+. For $Ni_{31}X$, the value of $\Delta\varepsilon'$ is much lower than the value obtained with NaX (cf. Table 1). As $\Delta\varepsilon'$ is linked to the number of cations which participate to the dielectric relaxation, this fact can be explained by 2 reasons :

i) the existence of 12 Ni^{2+} in S_I where they are probably blocked and therefore do not participate to the polarization.

ii) a local breakdown of the structure. Nevertheless, this effect is not the most important one because of the sensibility of the dielectric method to these phenomenon [8].

Samples reduced by hydrogen molecules

Sample A : the extent of reduction of Ni^{2+} corresponds to about 100 % [2]. The dielectric properties of this sample are very close to those obtained with NaX (cf. Table 1). This can be interpreted as follows :

- Firstly, they are no more Ni^{2+} in the structure.

- Secondly, the metallic particles are formed on the zeolite external surface [2] where they don't interact with the support.

Sample B : the extent of reduction is about 40 % in this case. The domain is shifted towards the low frequencies compared with the non-reduced Ni_8X sample. To explain this result we can consider 3 hypotheses :

- hypothesis 1 : the existence of some heterogeneity in the repartition of the non-reduced Ni^{2+} in the zeolite crystallites. The reduction is performed according to an intragranular diffusion process with the formation of a reaction front [7]. The reduced part of the crystallites (the periphery) contains mainly Na^+ cations which contribute to the high frequency part of the domain. On the other hand, the center of the crystallites contains the main number of the non-reduced Ni^{2+}. These cations have left the S_I sites [9]. So, they may modify the mobility of the neighbouring Na^+ cations much more than in the non-reduced Ni_8X.

- hypothesis 2 : the existence in the framework of small aggregates in the supercages (or clusters in the sodalite cages) having a residual electric charge. The existence of such species has been considered by some authors [10,11,12] in the case of Y zeolites.

- hypothesis 3 : as the reduction temperature was 573 K, the dehydroxylation process is not complete. So, there are protons in the structure near the tricoordinated alumina. In that case, it has been shown that the critical frequency is lowered [13].

At present, it is not possible to choose between the different hypotheses or to know whether all the effects exist simultaneously. Note, that with this type of sample there are Ni^+ cations in the framework [14]. But, as their number is very low, their dielectric effect is probably negligible.

Samples reduced by hydrogen atom beams

Various samples were reduced in the same conditions. Nevertheless, the dielectric spectra obtained were quite different. These various results could be ascribed to :

i) different extent of reduction.

ii) different repartition of the metallic aggregates in the samples.

iii) a local breakdown of the framework caused by microwave discharge.

We discuss here the results obtained with sample C which characterizes the most reproducible conditions. It has been reduced to about 100 % and contains aggregates homogeneously distributed in the lattice [3]. In comparison with sample A, the critical frequency is decreased by about a factor 12. This effect is analogous to the one observed with sample B. But, for sample C, we have to take only hypothesis 2 and hypothesis 3 into account. We can notice that the proton concentration is higher in sample C than in sample B. Indeed, in sample C, the extent of reduction is about 100 % and the dehydroxylation process doesn't exist at 293 K.

CONCLUSION

The results presented here show :

- the influence of the extent of exchange in Ni^{2+} on the dielectric properties of X zeolites.

- the homogeneity (or not) of the distribution of the cations in the different sites.

- the influence of metallic aggregates on the cation mobility. When these aggregates are sintered in the form of big particles at the zeolite external surface, the dielectric behaviour of the sample is the same as that of NaX.

- the local breakdown of the zeolite lattice during the reduction by hydrogen atom beams caused by microwave discharge.

So, the dielectric method gives a good contribution to the zeolite characterization.

REFERENCES

1 J.M. Wacrenier, J. Fontaine, A. Chapoton and A. Lebrun, Rev. Gen. Elec., 76 (1967) 719-725
2 M.F. Guilleux, D. Delafosse, G.A. Martin and J.A. Dalmon, J.C.S. Faraday I., 75 (1979) 165-171
3 M. Che, M. Richard and D. Olivier, J.C.S. Faraday I., 76 (1980) 1526-1534
4 P. Tabourier, J.C. Carru and J.M. Wacrenier, Zeolites (1982) (in press)
5 R. Freymann and M. Soutif, in La spectroscopie hertzienne appliquée à la chimie, Dunod, Paris, 1960, Ch. 4, p.36
6 G. Ravalitera, J.C. Carru and A. Chapoton, J.C.S. Faraday I., 73 (1977) 843-852 - Earlier papers are cited in this reference
7 M. Briend-Faure, J. Jeanjean, M. Kermarec and D. Delafosse, J.C.S. Faraday I., 74 (1978) 1538-1544
8 A. Chapoton, A. Lebrun and G. Ravalitera, C.R. Acad. Sc. Paris, série C, 271 (1970) 525-528
9 P. Gallezot, Y. Ben Taarit and B. Imelik, J. Phys. Chem., 77 (1973) 2556-2560
10 H. Beyer, P.A. Jacobs and J.B. Uytterhoeven, J.C.S. Faraday I., 72 (1976) 674-685
11 J.C. Vedrine, M. Dufaux, C. Naccache and B. Imelik, J.C.S. Faraday I., 74 (1978) 440-449
12 Ch. Minchev, V. Kanazirev, L. Kosova, V. Penchev, W. Gunsser and F. Schmidt in L.V.C Rees (Ed.), Proc. 5th Int. Conf. on zeolites, Naples, June 2-6, 1980, Heyden, London, 1980, pp. 355-363
13 L. Gengembre, J.C. Carru, A. Chapoton and B. Vandorpe in L.V.C. Rees (Ed.), Proc. 5th Int. Conf. on zeolites, Naples, June 2-6, 1980, Heyden, London, 1980, pp. 253-260
14 D. Olivier, M. Che and M. Richard, J. Phy. Chem. (1982) (in press).

ACKNOWLEDGMENTS

We thank Pr. Uytterhoeven and his group for helpful discussions we had about this subject.

MODIFICATION OF CHEMISORPTIVE AND CATALYTIC PROPERTIES OF Ni$^{\circ}$ HIGHLY DISPERSED ON ZEOLITES OF VARIOUS COMPOSITION.

G.N. SAUVION, M.F. GUILLEUX, J.F. TEMPERE and D. DELAFOSSE

ER 133, Réactivité de Surface et Structure, Université P. et M. Curie, 4 Place Jussieu, 75230 Paris Cedex 05, France.

ABSTRACT

It is shown that adsorptive and catalytic properties of well dispersed nickel metal on CeX zeolites can be considerably modified according to the support acidity. Results are discussed in terms of metal support interaction leading to the presence of hydrogen entities strongly chemisorbed on the metallic surface, originated during the reducing process.

INTRODUCTION

Numerous studies have shown the dependance of the reactivity of metal supported systems with the degree of dispersion of the metal and the strength of the metal-support interactions (ref. 1, 2). However, some controversies remain on the explanation of the modifications of reactivity of small metallic particles, particularly with metallic nickel.

In this paper, we compare the reactivity (adsorptive and catalytic properties) of small nickel particles of similar size distribution, supported on NaHCeX zeolites, differing mainly by their acidity.

EXPERIMENTAL

Three samples NiCeX were prepared by conventional exchange of a sodium X zeolite (sample III) or a partially decationated X zeolite (sample I and II) with O.1 mol.dm^{-3} solutions of Ni^{2+} and Ce^{3+} nitrates, successively. Samples composition, determined by chemical analysis, is given in table I. These samples are reduced at 623 K in a dynamic flow (5 dm^3.H^{-1}) of high purity hydrogen during 16 hours.

The metallic particle size distribution was determined by magnetic measurements using the Weiss extraction method (ref. 3) and ferromagnetic resonance study performed at 300 K, using a Varian Spectrometer (model C.S.E.-109-X band)

The hydrogen adsorption measurements were performed at room temperature and 100 Torr gas pressure in an apparatus equipped with a Texas Instruments Pressure

Gauge. Before adsorption measurements, the reduced samples were outgassed at 623 K. The temperature programmed desorption measurements were carried out on samples reduced at 623 K and cooled at room temperature under H_2. The nature and amounts of the gas evolved were analysed by mass spectroscopy.

The I.R. spectra were recorded on a Perkin-Elmer 580 13 spectrophotometer The measurements of the catalytic activity in butane hydrogenolysis were performed on a dynamic differential microreactor. The reagent gas, consisting of a ternary mixture of butane (pressure = 30 Torr), hydrogen (pressure = 350 Torr) and helium (pressure = 380 Torr), carefully purified and dried, were sent onto 40 mg of catalyst, at temperatures between 473 and 623 K. The gaseous reaction mixture was analyzed by gas chromatography with a flame ionization detector. Catalytic activities were referred to 523 K, the rates being expressed in turn over frequencies.

RESULTS AND DISCUSSION

The samples studied, when reduced at 623 K, lead, in all cases, to a well dispersed metallic state of nickel particles behaving as paramagnetic at 77 K as shown by magnetic measurements (Table I). The lineshape and line width observed by F.M.R. spectra indicated that for each sample, the distribution is homogeneous, the cristallite size increasing with the decreasing line width (ref. 4).

TABLE I

Characterization and reactivity of the samples.

Samples	Unit cell composition	α (a)	D(n.m) (b)	ΔH(G.) (c)	H/Ni^o (d)	N.10^3(sec^{-1}) (e)
I	$Ni_8Ce_5Na_{15}H_{40}X$	0.28	0.7	825	0	0
II	$Ni_{11}Ce_6Na_{22}H_{24}X$	0.73	≤ 1.5	610	0	0
III	$Ni_{11}Ce_{17}Na_9H_4X$	0.54	≤ 1.5	500	0.3	4

(a) Degree of reduction.
(b) D : particle size diameter in n.m determined by magnetic methods.
(c) F.M.R. line width in gauss.
(d) Hydrogen chemisorption measurements expressed by the (H/Ni^o) ratio obtained at 300 K.
(e) Butane hydrogenolysis rates expressed in turnover frequencies at 523 K.

The results of the adsorptive and catalytic properties study of these samples, versus the hydrogen chemisorption and butane hydrogenolysis reaction are summarized in Table I. It appears that the more acidic samples I and II display the same behaviour since they do not chemisorb hydrogen and have no hydrogenolysis activity, while sample III presents adsorptive and catalytic properties. However, one can note that, for this sample, the (H/Ni^o) ratio is less than those corresponding to the value of the metallic dispersion and that the hydrogenolysis activity is not very high. These results have led us to the study of the state of the metallic surface by T.P.D. analysis and infra-red spectroscopy.

The infra-red spectroscopy has evidenced the existence, for all the samples, of a band located at 1990 cm^{-1}, which may be ascribed to an hydrogen metal bond (ref. 5). This band, more intense in the case of samples I and II, disappears by evacuation at 623 K.

The T.P.D. spectra of the samples studied are shown in Fig. 1. It is observed, in all cases, the removal of water and hydrogen till high temperatures.

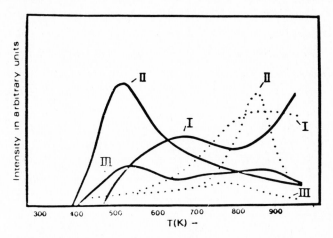

Fig. 1. Thermodesorption spectra showing the removal of H_2O (———) and H_2 (··········) from sample I, II and III.

The disappearance of the I.R. band at 1990 cm^{-1} during desorption at 623 K seems to be correlated to the removal of water observed by T.P.D. analysis in the same temperature range.

Furthermore, the large quantities of water observed for sample II can not only originate from the desorption of OH groups remaining on the surface after the reduction process at 623 K. Therefore, it can be postulated that, during desorption in this temperature range, an hydrogen species dissociated on the metal leaves the surface as water combining with neighbouring OH groups of the support (ref. 6).

At higher temperatures, desorption of molecular hydrogen can be observed with significantly larger quantities of this gas evolved for sample I and II. Therefore, one can understand why samples with high percentage of nickel particles covered with hydrogen chemisorbed during the reduction process, will no longer chemisorb molecular hydrogen nor possess catalytic activity versus hydrogenolysis reaction. So, the adsorptive and catalytic properties of the samples studied are highly dependent on the hydrogen coverage of metallic nickel particles. Attempts to see the eventual appearance of some catalytic activity of samples I and II, after thermodesorption treatment till 923 K, in order to remove chemisorbed hydrogen, have been carried out and proved unsuccessfull.

F.M.R. measurements recorded during this thermodesorption treatment have shown that the three samples behave differently. Sample I is progressively oxidised till a complete disappearance of the F.M.R. signal and consequently a total oxidation. The lack of hydrogenolysis activity of this sample is obviously correlated to the oxidation of the metallic nickel. Sample II and III are particularly stable in size since, for both samples, the line-width of the respective spectra remains unchanged, meanwhile the intensity of the signal increases and this, to a much greater extent for sample II. This can be explained by the increase of the magnetization of the samples due to the desorption of the chemisorbed hydrogen. However, contrarly to sample III, sample II cannot catalyse hydrogenolysis as mentioned above. In order to elucidate the cause of the inactivity of sample II, a hydrogen flow is sent, between 473 and 623 K onto the catalyst thermodesorbed at 923 K. It was observed that hydrogen is again strongly chemisorbed and leaves the metallic surface at high temperatures as in the case of hydrogen retained on the nickel during the reduction process. So, one can understand that during the hydrogenolysis run, on the bare nickel particles, hydrogen reuptake, from the reagents mixture, occurs and consequently suppresses the hydrogenolysis activity.

So, the adsorptive and catalytic behaviours of the three samples studied seem to be highly dependent on the hydrogen coverage of metallic nickel particles. Furthermore, it appears that higher hydrogen coverage is observed for the more acidic samples I and II.

In the case of sample I, the more acidic, small nickel cristallites are stable only in presence of adsorbed hydrogen (ref. 7) and undergo a progressive oxidation when hydrogen is desorbed. Nickel particles dispersed on sample II, with lesser acidity than sample I, are stable, don't oxidise by hydrogen desorption, but keep too much hydrogen at 523 K to be reactive.

It is only with sample III which presents the lesser acidic properties, and retains hydrogen in small quantities after the reduction process as proved by T.P.D. experiments, that adsorptive and catalytic properties for hydrogenolysis can be observed.

All the above results strongly suggest the indirect influence of the acidity of the support on the nickel particles reactivity.

The presence of electron acceptor sites on the support (Lewis acid sites) can explain that some electron depletion occurs at the metallic surface of small crystallites (ref. 8) leading to strong metal-support interactions. It seems likely that the greater the acidity of the support, the greater the electron deficient character of the metal. Such an electron depletion of the metal can enhance its affinity to chemisorb electron donating molecules, such as hydrogen, during the reducing process (ref. 9). So, small nickel particles, when in strong interaction with highly acidic support retain hydrogen till high temperatures, and as a result, adsorptive and catalytic properties (in hydrogenolysis reaction) of these particles are suppressed.

Generally, in the literature, the modifications of chemical properties of highly dispersed metal have been ascribed either to the loss of collective properties of the metal, or to the existence of strong metal-support interactions, modifying the electronic structure of the metal (ref. 10). The present results seem to show that adsorptive and catalytic properties of small nickel particles are dependent on possible existence of strong metal support interactions which are able to determine indirectly the coverage of the metal surface by hydrogen species chemisorbed during the reducing process.

Acknowledgments

The authors are grateful to Dr Djega-Mariadassou for assistance with the temperature-programmed desorption study.

234

REFERENCES

1   P.G. Menon and G.F. Froment, Applied catalysis, 1 (1981) 31-48.
2   P. Gallezot, Catal. Rev -Sci. Eng., 20 (1) (1979) 121-154.
3   P. Weiss and R. Forrer, Ann. Phys. (Paris), 5, (1926) 153.
4   D. Olivier, M. Richard, L. Bonneviot and M. Che, Studies in surface science and catalysis, Vol. 4, Elsevier, Amsterdam, (1980) 193-199.
5   T. Nakata, J. of chem. Phys., 65 (1976) 487-488.
6   T.M. Apple, P. Gajardo and C. Dybowski, J. Catal., 68 (1981) 103.
7   G. Martino, Studies in surface science and catalysis, Vol. 4, Elsevier, Amsterdam, (1980) 399-413.
8. T.M. Tri, J. Massardier, P. Gallezot and B. Imelik, Proc. of the 7th Int. Cong. on catalysis, part A, Tokyo 1980, Elsevier, Amsterdam (1981) 266-275.
9   R.P. Messmer, S.K. Knudson, K.H. Johnson, J.B. Diamond and C.Y. Yang, Phys. Rev., B 13 (1976) 1396-1415.
10 S.J. Tauster, S.C. Fung and R.L. Garten, J. Amer. Chem. Soc., 100 (1978) 170-175.

\

NICKEL IN MORDENITES - FORMATION AND ACTIVITY OF METALLIC COMPLEXES, CLUSTERS AND PARTICLES

E.D. GARBOWSKI[*], C. MIRODATOS, M. PRIMET

Institut de Recherches sur la Catalyse, C.N.R.S.
2, avenue Albert Einstein

69626 - VILLEURBANNE CEDEX -

[*] U.E.R. Chimie Biochimie Université Claude BERNARD - LYON I -

ABSTRACT

Nickel ions have been exchanged in soded and protonated mordenite. IR and UV spectroscopies and magnetic methods show that dehydrated samples have nickel ions located in cationic sites of low symmetry with a tendency to agglomerate in protonated mordenite. These ions are able to coordinate CO and NO at room temperature for forming stable complexes. Nitrosyl complexes react with hydrogen at moderate temperature giving N_2O, N_2 and H_2O. Carbonyl complexes may be reduced at low temperature into Ni(I) species and Ni(O) ultradispersed clusters and small particles. These species are not active for methanation. At higher temperature, sintering of nickel occurs and large particles of metal (surrounded by nickel oxide in protonated mordenite) are formed outside the zeolite.

INTRODUCTION

Extensive studies have been devoted to metallic phase and metallic clusters embedded in zeolite with the ultimate goal to prepare highly dispersed and stable metal particles (1, 2). In this way, various investigations and approaches in ion chemistry - coordination state, location in cationic sites, microstructure inside zeolites, reactivity towards gases as NO, CO, H_2 are clearly required to understand, then to control the formation of these metal particles.

We report in this work some aspects of that question by following the changes of nickel state in sodium and hydrogen mordenite, when ligands as CO or NO are added, during reduction and catalysis.

EXPERIMENTAL

The sodium form of mordenite zeolite which is referred to as NaM was provided by NORTON (ZEOLON 900). The decationated form was prepared by contacting the sodium form with a saturated NH_4Cl solution. Nickel exchanges were performed at room temperature with 0.5 M nickel nitrate solutions. Catalysts were activated (deammination, dehydration) at 500°C under flowing dry oxygen, then evacuated at the same temperature. Chemical analysis gave the following unit cell composition.

NiNaM : $Ni_{2.2}$ $Na_{3.4}$ $H_{0.2}$ $(AlO_2)_8$ $(SiO_2)_{40}$

NiHM : $Ni_{1.8}$ $Na_{0.0}$ $H_{4.4}$ $(AlO_2)_8$ $(SiO_2)_{40}$

IR spectra were recorded on small pellets of 10-15 mg with a P.E. 580 or P.E. 125 spectrophotometer. UV measurements were performed with an optica MILANO CF_4DR diffuse reflectance spectrometer. Magnetic measurements allowed to determine some aspects of nickel morphology using LANGEVIN or WEISS method depending the samples were superparamagnetic or not (3, 4). X-ray diffraction pattern were recorded continuously under controlled hydrogen atmosphere at programmed temperature in order to get information (temperature reduction, particle growth) on nickel reduction (5).

Kinetic experiments were carried out in a flow system with a fixed-bed reactor at atmospheric pressure. Gas analyses were performed by gas chromatography with catharometric and flame-ionization detectors.

Results and discussion

a) Dehydrated samples

After O_2 and vacuum treatment at 500°C, the UV spectra of both NiNaM and NiHM mordenite show that Ni^{2+} ions are free from water and coordinated to oxygen ions of the zeolite. Number and position of the bands -425, 470, 570, 870, 1600, and 1820nm-suggest that the obtained complexes are highly asymmetrical. They would be located in the low symmetry wall sites along the main channels, which are referred to as sites IV and VI in the MORTIER'S nomenclature (6,7).

Although the UV spectra of NiHM and NiNaM are quite similar, a marked discrepency is revealed by magnetic measurements. For NiNaM, the experimental magnetic susceptibility clearly indicates that nickel ions are non interacting i.e. isolated in the zeolitic mattrix. Moreover, the experimental value of the magnetic moment (μ = 3.24 BM) slightly higher than the theoretical value of the Ni^{2+} free ion reinforces the idea of a strongly distorted coordination state. Differently, an unusual high susceptibility is measured for NiHM sample, (10 fold the free Ni^{2+} susceptibility at 77 K). This has been assigned to a cooperative phenomenon which suggests that the nickel ions can agglomerate along the channels and therefore be structured in nickel-oxide-like small particles (8).

b) CO adsorption

Absorption of carbon monoxide (20Torr) at room temperature causes pronounced changes of the UV spectrum. The three remaining bands at 425, 800 and 1630 nm suggest that a new carbonyl complex with an octahedral symmetry has been formed (Fig. 1a). Infrared spectrum shows a ν_{CO} band at 2205 cm^{-1} whose intensity varies with the CO pressure. It has been checked that the same band is not observed in the case of CO adsorbed on nickel free zeolite samples. This frequency has been assigned to carbon monoxide interacting with accessible Ni^{2+} ions (9, 10). No major change is noted in magnetic measurements.

c) CO + H₂ adsorption

Reactivity of the carbonyl complexes towards hydrogen is tested by contacting the zeolite samples with a CO + H_2 mixture (CO/H_2 = 1/4) at increasing temperature.

-NiNaM

For NiNaM sample, reaction occurs at room temperature. Two IR bands at 2130 and 2080 cm develop in addition of the 2205 cm^{-1} one (Fig. 2b). The intensity of both bands increases with the CO pressure but in a constant ratio. The 2130 and 2080 cm^{-1} absorptions are removed by vacuum treatment at room temperature. The magnetization recorded in similar conditions (flowing H_2 + CO mixture instead of static contact) is decreased by comparison with the signal of dehydrated NiNaM sample. This gives a magnetic moment of 2.5 BM and a calculated number of unpaired electron per nickel ion of 1.8.

Previous studies mainly in Y zeolites (11) have shown that Ni^{2+} ions can be reduced into Ni^+ species by hydrogen treatments at moderate temperature. IR spectroscopy and magnetic measurements clearly indicate that we are dealing with the same process for NiNaM, i.e. a partial reduction of Ni^{2+} into Ni^+ ions, and more precisely into gem dicarbonyl Ni^+ $(CO)_2$ according to the admitted assignment of the 2130 and 2080 cm^{-1} bands (9, 10). Although the UV absorption of Ni(II) carbonyl species renders the observation difficult, a slight change around 700 nm on the UV spectrum agrees with this formation of Ni^+ ions whose internal transition is generally expected at 740 nm. Moreover, EPR measurements performed at 77 K reveal a paramagnetic species : a weak orthorhombic signal was obtained as soon as CO was admitted. In the same time, 2 UV bands appear and grow at 375 and 315 nm, especially if temperature is raised up to about 150°C (Fig. 1, b-e). At fixed temperature, these bands develop following a kinetic law which can be expressed as an Elovich equation. F(R) = a + Logt where F(R) is the KUBELKA MUNK function of the band reflectance. This equation has already been reported in the literature for instance to account for the reduction of NiO by H_2 (12). Refering to the works of MOSKOVITS and HULSE (13) and KOLTZBUCHER and OZIN (14) which deal with optical spectra of nickel atoms or cluster embedded in argon mattrix, and taking into account the large differences between their experimental conditions and ours, we tentatively assign the two bands at 375 and 315 nm to Ni(O) very small clusters formed by the reduction of the carbonyl complexes (7). IR spectroscopy considerably supports that scope. Heating at 100°C then at 200°C the CO + H_2 mixture causes reduction of Ni^{2+} ions since the intensity of the 2205 cm^{-1} band decreases to a large extent (Fig. 2, c-d). In the same time, the

238

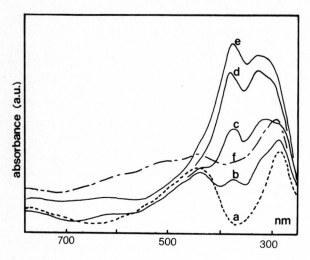

Fig. 1. UV spectra of NiNaZ contacted with CO at room temperature (a); CO + H_2 for 15min.(b), for 20 h (c), at 100°C (d), 150°C (e), evacuated and contacted with oxygen (f).

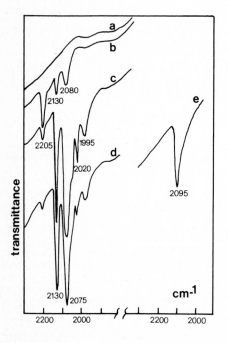

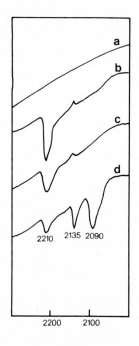

Fig. 2. IR spectra of NiNaZ, contacted with O_2 and evacuated at 500°C (a); with CO + H_2 at 25°C (b); at 100°C for 16 h (c),at 200°C for 2 h (d); evacuated at 25°C (e).

Fig. 3. IR spectra of NiHZ contacted with O_2 and evacuated at 500°C (a), with CO + H_2 at 25°C (b), at 100°C for 14 h (c); at 200°C for 3 h (d).

band at 2080 cm^{-1} develops considerably, much more faster than the 2130 one which even decreases at 200°C. In addition, weak bands at 2020 and 1980 cm^{-1} are observed with a maximum intensity around 100°C. When the sample is desorbed at 25°C, all the ν_{CO} bands previously observed disappear except the 2080 cm^{-1} one which shifts to 2095 cm^{-1} (Fig. 2e). The latter desappears by treating the sample under oxygen. This band which partly overlaps the 2080 cm^{-1} band of the dicarbonyl Ni$^+$(CO)$_2$ species is interpreted as carbon monoxide mono adsorbed on metallic nickel. The low frequency of the band and the very easy reoxidation by oxygen suggest that we are dealing with very small particles. Concerning the doublet at 2020 - 1980 cm^{-1} two assignments are available. On one hand, it can be also interpreted in terms of metallic nickel formation ; the metal would be in the form of microclusters able to chemisorb CO according to MOSKOVITS et al. (15). On the other hand, it may indicate the formation of new Ni(I) species which could be the dinuclear double-bridged entity (Ni(CO)$^+$)$_2$ as postulated in NiCaX zeolites by BOSON-VERDURAZ et al (10). The formation of Ni$^+$ carbonyl species could come from the reverse disproportionation reaction Ni^o + Ni^{2+} $\rightleftharpoons$ 2N$_2^+$ as postulated by KASAI et al (9). In both cases, theses species are quite unstable since they disappear by pumping at room temperature or when temperature of the CO + H$_2$ mixture is raised to 200°C. After the same treatment at 100°C in CO + H$_2$ flowing conditions, no bulk metallic nickel is detected by magnetism, which is quite consistent with the picture of very small particles and/or clusters covered with linear carbon monoxide. When the NiNaM sample is heated at higher temperature under CO + H$_2$ (from 200 to 400°C), UV spectroscopy-growing background - and magnetism - fast increase of magnetization - clearly show that bulk metal is formed while carbonyl species are destroyed. Moreover, the two IR bands which develop at 3020 and 1305 cm^{-1} point out that methane is formed at around 400°C.

-NiHM

The effects of CO + H$_2$ mixture on hydrogen mordenite are much more limited.

UV spectrum indicates that most of nickel ions do not react with hydrogen since no Ni$^+$ species and only a small doublet at 375 and 315 nm (atomic nickel) is detected. Similarly the IR spectrum displays only one band at 2205 cm^{-1}, i.e. CO-Ni^{2+} (Fig. 3b). A still high value of nickel susceptibility measured at 77 K suggests that a large part of nickel ions remain agglomerated inside the zeolite channels. When temperature is raised to 200°C, some evidence of Ni$^+$ is provided by IR spectroscopy with a slight decrease of the Ni^{2+} - CO band but in no case metallic particles or microclusters are detected (Fig. 3 c, d). At 400°C, a background characteristic of bulk metal starts to develop in the UV spectrum but at a much smaller extent than in the case of sodium mordenite. This metal phase appears to be inactive since no methane is detected by infrared.

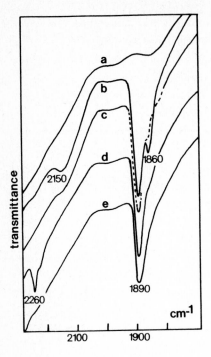

Fig. 4 : IR spectra of NiNaZ contacted with O_2 and evacuated at 500°C (a) ; with NO (10 Torr) at 25°C (b) ; evacuated at 25°C (c) ; with H_2 (100 Torr) at 100°C (d) ; at 200°C for 3 h (e).

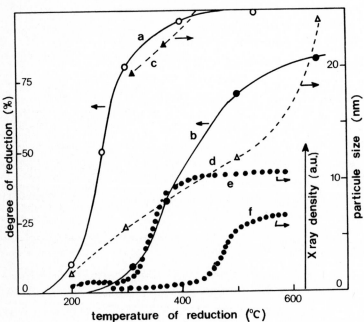

Fig. 5 : Degree of reduction (a, b), average nickel particle size (c, d) and nickel 111 X-ray density (e, f) as a function of temperature of reduction for NiNaM (a, d, e) and NiHM (b, c, f) samples.

d) NO adsorption

NO adsorption onto NiNaM zeolite leads to the formation of nitrosyl complexes whose geometry would be bipyramidal trigonal, according to the intensity and number of bands observed in UV-reflectance spectroscopy (7). IR spectroscopy suggests that in addition to the main species Ni^{2+}-NO which absorb at 1890 cm^{-1}, the nitrosyl complexes could also be in the form $Ni^{+}NO^{+}$ (2150 cm^{-1}) and $Ni^{2+}(NO)_2$ (1860 cm^{-1}) (Fig. 4b). Reaction with H_2 leads to the following phenomena :

- at moderate temperature ($\leqslant$ 200°C) NO adsorbed is reduced into N_2O (IR band at 2260 cm^{-1}, Fig. 4d) water (IR and UV bands) and OH groups.

- at higher temperature, the NO - Ni^{2+} IR band completely vanishs, N_2O is in turn reduced into gazeous nitrogen and a UV background of bulk nickel develops.

No clue of metal microstructure formation is provided by these experiments ; due to the rather high stability of nickel nitrosyl complexes, the nickel ions are reduced (when they are free of their NO ligand) at temperatures high enough to allow the metallic phase to sinter. Magnetic measurements well indicate that as soon as 300°C, reduction in Ni-NO mixture leads to particles larger than 10 nm. Reduction is never completed, even at temperatures higher than 500°C. The same observations hold for NiHM sample.

e) Reduction in hydrogen

Nickel reduction is performed after dehydration at 500°C at increasing temperature either in static atmosphere (100 Torr H_2 pressure) or in flowing hydrogen. No major discrepencies are evidenced between the two methods. Figure 5 gives for both zeolites the degree of reduction and average metal particle size measured by magnetism, and the intensity of the metallic nickel 111 X-ray as a function of the reduction temperature (for a given reaction time of 15 h). A temperature shift is observed for the detection of metal between magnetic and X-ray measurements. It comes probably from the lower sensibility of X-rays, which detect only particles large enough to allow the diffraction. These data are totally consistent with the informations given by UV spectrometry (disappearence of Ni^{2+} bands and increase of metal nickel background).

Reduction of Ni(II) ions in sodium mordenite is shown without ambiguity to be much more easier than in hydrogen mordenite, for which it has never been completed ($\leqslant$ 85 %), as recently noted by SUZUKI et al (16). Moreover, the sintering of nickel particle is considerably enhanced in NiHM sample. For the latter zeolite, hydrogen volumetric measurements and magnetism reveal that bulk metal particle display only a very small metallic surface. It would be mainly a shell of nickel oxide around a core of metal. It is shown elsewhere (17) that the only way to gain some stable metallic surface is to reduce under H_2 pressure after a steaming treatment decreasing partially the acidity of the catalyst. This poor reducibility of nickel ions in hydrogen mordenite is certainly to be related to the concentration of zeolite hydroxyl groups which favors the oxidative reaction (7, 8, 16).

$$Ni^{\circ} + 2OH_{zeol.} \rightleftarrows Ni^{2+} + 2O^-_{zeol.} + H_2$$

In the same way, the role played by OH groups explains the uncomplete reduction observed under a NO + H_2 mixture, even for NiNaM zeolite because of the formation of OH groups when NO is reduced into NiO and N_2.

f) Catalytic activity

Both catalysts NiNaM and NiHM are tested in methanation reaction. After standard activation treatment, the CO + 4 H_2 mixture is directly admitted without prereduction. Significant activity is detected for reaction temperature higher than 250°C for NiNaM and 350°C for NiHM mordenites. At 400°C, catalytic activities of NiNaM and NiHM are respectively 11 x 10^9 and 3 x 10^9 molec./sec x cm^2 (note that NiHM specific activity is a rough evaluation due to its near negligible metallic surface). Apparent activation energy for NiNaM and NiHM is around 28 kcal mole^{-1}. In both cases, selectivity in methane is higher than 90 %, as expected when methanation is performed at high temperature (18).

From these data we will emphasize the following points :

- methanation seems to occur upon the same mechanism for the two zeolites (similar activation energy and specific activity)

- in both cases, the temperature at which methane begins to be significantly detected corresponds rather well to the formation of large metal particles (> 3 nm) This reinforces our previous IR and UV observations that the very small particles or clusters of atoms formed at low temperature under H_2 + CO mixture are not active for methanation. One possible explanation is that the latter particles are too small to allow the methane formation, owing to the high number of adjacent nickel atoms required for methanation according to DALMON et al (18). An other point is that the experimental values of catalytic activity are much lower than the values given in similar conditions for non zeolitic catalysts as Ni/SiO$_2$ (18) Previous works have shown that the most efficient supports for methanation are the less acidic (MgO > Al$_2$O$_3$ > SiO$_2$). In our case, the mordenite framework, and especially the hydrogen one could act as a powerful electron drawing, then lower the electronic density of metal. That would hinder the adsorption of the electrophilic reactant CO, then the formation of methane.

CONCLUSIONS

It has been shown in this study that mordenite behaves as a peculiar support for nickel. In general, nickel ions are in a coordinence state of low symmetry. They are well dispersed when the other compensating cations are mainly sodium but have a tendency to agglomerate in the hydrogen mordenite. These ions can form carbonyl or nitrosyl complexes inside the zeolitic channels. Depending the absence or presence of OH groups in the framework, these complexes are more or less reduced by hydrogen forming new clusters in the case of carbonyl. Thus, well dehydrated sodium mordenite can stabilize atoms in ultra dispersed state, preventing sintering as far as the reduction temperature is mild. Unfortunatly, these metallic species are not active in methanation. Other kind of catalytic reaction as nitric monoxide reduction would better suit this catalyst. When reduced at higher temperature, nickel forms large particles outside the zeolite (much more easily for protonated mordenite, although the metallic particles remain surrounded by nickel oxide) which could be in electronic interaction with the support.

Acknowledgments : The authors thank Dr. G. Bergeret for helpful X-ray experiments and discussions.

REFERENCES

1. Catalysis by zeolites : Studies in Surface Science and Catalysis 1980, Imelik Ed.
2. J. Rabo in "Zeolite chemistry & catalysis " (Ed. J.A. RABO) ACS monograph, Washington 1976.
3. P.W. Selwood "Chemisorption and magnetization" Academic Press, New York, 1975.
4. M. Primet, J.A. Dalmon, G.A. Martin, J. Catalysis 1977, 46, 25.
5. G. Bergeret, P. Gallezot, B. Imelik, J. Phys. Chem., 1981, 85, 412.
6. W.J. Mortier, J. Phys. Chem. 1977, 81, 1334.
7. E.D. Garbowski, C. Mirodatos, M. Primet and M.V. Mathieu (submitted for publication).
8. C. Mirodatos, J.A. Dalmon, E.D. Garbowski, D. Barthomeuf, "Zeolites". In press.
9. P.H. Kasai, R.J. Bishop, D. Mc Leod, J. Phys. Chem., 1978, 82, 279.
10. F. Bozon-Verduraz, D. Olivier, M. Richard, M. Kermarec, M. Che, in Press
11. E. Garbowski, M. Primet and M.V. Mathieu, A.C.S. Symposium Series Molecular Sieves II, 1977, 40, 281.
12. A.M. Peers, J. Catal., 1965, 4, 672.
13. M. Moskovits and J.E. Hulse, J. Chem. Phys. 1977, 66, 3988.
14. W. Klotzbücher and G.A. Ozin, Inorg. Chem. 1976, 15, 292.
15. J.E. Hulse and M. Moskovits, Surface Science, 1976, 57, 125.
16. M. Suzuki, K. Tsutsumi, H. Takahashi, "Zeolite" 1982, 2, 51.
17. C. Mirodatos, J.A. Dalmon, D. Barthomeuf, to be published.
18. J.A. Dalmon, G.A. Martin, Proceedings 7th Int. Cong. Catalysis, Tokyo, 1980, 402.

DISPERSION OF NICKEL AND RUTHENIUM IN ZEOLITES L, Y, AND MORDENITE

S. NARAYANAN

Catalysis Section, Regional Research Laboratory, Hyderabad 500009, India

ABSTRACT

Nickel and ruthenium zeolites are compared for their reducibility and benzene hydrogenation activity. Ruthenium is better dispersed than nickel under the same pretreatment conditions. Calcination in air prior to reduction, decreases the ruthenium area via the formation of large crystallites. Turnover number increases with the available ruthenium surface and is independent of the zeolite type whereas in nickel system the support seems to play a part. Crystallite sizes calculated from hydrogen adsorption and XRD are compared. XPS and ESR of 3.6% ion exchanged NiM reveal the presence of more than one species of nickel, viz., Ni(2+), Ni(1+) or probably Ni(OH)$_2$.

INTRODUCTION

Characterizations of dispersed supported metal catalysts are important in any catalytic reaction where the activity and selectivity depend on dispersion and crystallite size. The specific methanation activity of nickel increased with increasing crystallite size and the existence of an optimum crystallite size has been suggested (ref.1,2). The particle size of ruthenium seems to influence the hydrocarbon chain length in FT synthesis (ref.3).

Hydrogen adsorption has been recommended for the measurement of ruthenium and nickel surface areas (ref.4). However, different authors (ref.5-12) have taken the hydrogen adsorbed at different equilibrium pressures as the monolayer volumes in their calculations of metal areas and there is no uniformity in the selection of the isotherm temperature. Eventhough, the same technique was used for measuring metal areas in nickel and ruthenium zeolites (ref.11,12), the applicability of this method is a subject of discussion. Because of their importance, nickel and ruthenium zeolites are chosen for this study. The effect of support on reducibility, dispersion, catalytic activity and the validity of adsorption measurements for crystallite size calculations are discussed.

EXPERIMENTAL

Catalyst preparation

Nickel zeolites were prepared both by impregnation and by ion exchange methods. RuL and ruthenium mordenite (RuM) were prepared from aqueous RuCl$_3$ solution whereas RuY and RuX were hexaammine ruthenium. Sodium and nickel were analyzed by atomic absorption and ruthenium by UV method (ref.12). The catalysts were pelletized without binder and particles of 1200-1600 μm were used for adsorption and catalysis.

Adsorption studies

Adsorption measurements were made using a constant volume system. Fresh catalyst was calcined at 723 K in oxygen for 2 h followed by reduction in hydrogen for 6 h. After evacuation for a few minutes at the reduction temperature, the catalyst was degassed at 633 K for 2 h or till a pressure of 10^{-5} Torr was achieved. Dead space was determined using helium at 303 K. It was further evacuated at 633 K for 2 h before the isotherm measurements were made at room temperature. Doses of hydrogen were added at 5-10 minutes intervals till a pressure of 300 Torr was reached. Measurements were repeated after evacuation at the same temperature for 30 minutes to remove any physisorbed gases.

The straight line parallel isotherms (see Fig.1) were extrapolated to zero pressure. Hydrogen adsorbed at zero, 100 and 200 Torr and also the difference between the two isotherms were taken for calculation of the average size of metal crystallite from the relation,

$$d(nm) = 6.10^3/ P.S \qquad (1)$$

where d is the crystallite size, in nm, S is the surface area of metal in m^2g^{-1} and P is the density of metal which is 8.9 $g.cm^{-3}$ for nickel and 12.06 $g.cm^{-3}$ for ruthenium. The hydrogen adsorption data are given in Table 1.

X-ray analysis

X-ray diffraction measurements were performed on samples that had been previously pretreated and exposed to atmosphere. Ni(111) peak was used for line broadening calculations. Crystallite size was calculated using Scherrer equation (ref.13).

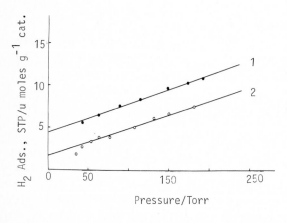

Fig.1

Typical adsorption isotherm

1. adsorption at RT
2. adsorption at RT after evacuation for 30 mins at RT

Catalysis experiment

The experimental set up has been described in detail earlier (ref.12). Benzene hydrogenation was carried out in the vapour phase. The pretreated catalysts were activated in flowing hydrogen for at least 2h at 673 K before each experiment. By using a calibrated motorized syringe benzene was introduced and the liquid products were analyzed by gas chromatograph.

XPS measurements

Vacuum generators ESCA 3MK II with AlK_α x-ray source ($h\nu$ = 1486.6 ev) was used. A slit width of 4 mm and analyzer energy of 50 ev were employed. The samples were mounted onto the probe and were evacuated at 473 K for 2 h and then were transferred to the analyzer chamber without exposure to atmosphere.

RESULTS AND DISCUSSION

Adsorption studies

The straight line isotherms (see Fig. 1) indicate either the slow attainment of equilibrium or some physisorption or both. The monolayer volume can be taken by extrapolating the first adsorption to zero pressure, assuming this will compensate for the physical adsorption (ref.10). Better still would be to use the difference between the two isotherms 1 and 2 on the assumption that evacuation for 30 minutes at room temperature would not remove the chemisorbed hydrogen. This method could be useful when there is a possibility of weak adsorption of hydrogen and the support is as porous as zeolites having high surface area.

The crystallites calculated by the above procedures were 4-10 times larger when compared with those by XRD measurements (Table 1, Column 1 and 2). Such a large difference in crystallite size between techniques have been reported (ref.14,15). In the case of nickel-zeolon, Brooks and Christopher (ref.11) determined the metal crystallites by hydrogen adsorption and by XRD. They varied as much as 200-300%. However, in the present study, if the hydrogen adsorbed at 200 Torr was taken as the monolayer volume, then the crystallite size calculated agreed quite well with XRD measurements (Table 1). This does not preclude the importance of the 1st method which is based on the assumption that the chemisorbed hydrogen could not be removed by evacuation at room temperature for 30 minutes. Probably if enough time is given for the attainment of equilibrium after each addition of hydrogen and the difference in the two isotherms is then taken as a monolayer volume, one might get crystallite sizes agreeing to XRD.

The dispersion in nickel zeolites is not good. Impregnated catalysts, NiL in particluar, have a relatively better dispersion compared to the ion exchanged zeolites. It is difficult to say, for sure, whether the calculated large crystallites and poor dispersion are the result of incomplete reduction or poor adsorp-

TABLE 1

Hydrogen adsorption data

Catalyst	Method of preparation	Metal/wt %	H$_2$ ads, STP/ μ moles g^{-1} cat				Metal area/ m^2.g^{-1} metal				Dispersion/%				Crystallites/nm					TON/ s^{-1}.10^{-3}
			1	2	3	4	1	2	3	4	1	2	3	4	1	2	3	4	XRD	
3-NiM	IE	3.6	2.8	4.2	7.7	11.4	6.1	9.1	16.7	24.7	0.9	1.4	2.5	3.7	110	74	40	27	29	1.9
5-NiM	IE	3.0	1.3	3.1	5.6	8.0	3.4	8.1	14.6	21.0	0.5	1.2	2.2	3.1	198	83	46	32	29	1.9
7-NiM	IE	2.0	2.1	5.1	7.1	9.3	8.2	19.9	27.7	36.5	1.2	3.0	4.2	5.4	82	34	24	18	23	1.6
9-NiM	IE	2.4	0.9	2.6	4.9	7.2	2.9	8.5	16.0	23.3	0.4	1.3	2.4	3.5	232	79	42	29	23	1.9
10-NiM	IE	3.5	1.4	4.4	7.5	10.9	3.0	9.8	16.8	24.3	0.5	1.5	2.9	3.6	224	69	40	28	29	1.3
12-NiM	IMP	3.7	4.0	8.0	12.0	16.4	8.5	17.0	25.6	35.0	1.3	2.6	3.8	5.2	79	40	26	19	--	1.3
13-NiY	IMP	3.7	1.8	2.0	5.7	9.4	3.9	4.3	12.2	20.0	0.6	0.6	1.8	3.0	175	157	55	34	--	37.0
14-NiL	IMP	3.5	10.2	15.0	18.8	22.8	22.8	33.5	42.0	51.0	3.4	5.3	6.3	7.6	29	20	16	13	--	27.6
15-RuM	IMP	2.4	2.5	3.0	7.7	12.5	8.3	9.9	25.5	41.4	2.1	2.6	7.2	10.7	60	50	19	12	--	23.0
16-RuL	IMP	0.9	3.2	4.3	7.4	10.6	27.8	37.4	64.3	92.1	7.2	9.6	16.6	23.8	18	13	8	5	--	47.6

1. Difference between two isotherms,   2. At zero pressure,   3. At 100 Torr,   4. At 200 Torr
Benzene hydrogenation at T = 413 K, LHSV = 0.33, H/HC = 0.33, Dispersion, Crystallite size and
TON = calculated using hydrogen adsorbed at 200 Torr.   IE = Ion exchanged,   IMP = Impregnated

tion, which can be due to (i) non availability of nickel to hydrogen (ii) un-
certainties as to the fraction of metal surface in contact with support and
therefore, inaccessible to hydrogen. Calcination of these catalysts in oxygen
prior to reduction would have probably affected the reducibility as was observed
on nickel-alumina (ref.10). It could also be possible in the case of nickel
mordenite that the metal particles inside the cavities of the zeolites inter-
acted strongly with the lattice and were not completely reduced. But this could
not explain the large crystallites measured by XRD.

It has been reported (ref.16-18) that nickel sintered easily forming large
metal particles at the external surface of zeolites. The reduction of nickel in
zeolites is more complex than of other transition metals or polycationic metals
such as ruthenium (ref.19-22). Ruthenium is better dispersed than nickel especially if it is reduced without prior calcination in oxygen. This is clear
from the comparison of metal areas, dispersion and crystallite size, reported
in Tables 1 and 2.

TABLE 2
Influence of pretreatment conditions on dispersion and crystallites size of
ruthenium.

Catalyst	Metal/ wt %	Pretreatment conditions					
		2h O_2 + 6h H_2 at 723 K			6h H_2 at 723 K		
		Disp/%	Metal area/ m^2g^{-1}Ru	Crystal- lite/nm	Disp/%	Metal area/ m^2g^{-1}Ru	Crystal- lite/nm
RuL	0.9	24	92	5	100	537	0.9
RuM	2.4	11	41	12	60	313	1.5

Catalysis studies

Hydrogenation of benzene was used as model reaction to compare the catalytic
activity (Table 1). Among the ion exchanged catalysts, there is not much difference in dispersion or the crystallite size or the catalytic activity. Of the
impregnated catalysts, NiY and NiL are better dispersed compared to NiM and have
higher catalytic activity by an order of magnitude. Dispersion follows the order
NiY< NiM< NiL whereas the activity follows the order NiY >NiL >NiM. It looks as
though the activity is not directly dependent on dispersion alone and the zeolite is suspected to play a part in it.

Ruthenium mordenite is not so active compared to RuL. With the least metal content, RuL is the most active of all the catalysts. Ruthenium zeolites pretreated

250

without prior calcination in oxygen, were also compared for dispersion and benzene hydrogenation activity (see Fig.2). It is seen that with the increase in metal content, there is a smooth decrease in the dispersion and an increase in the crystallite size. The turnover number decreases smoothly with the metal content whereas it increases in a linear fashion with the metal surface area of ruthenium. This shows, contrary to the observation on nickel zeolites, that the hydrogenation is directly related to dispersion and the available ruthenium surface and is independent of the zeolite type. The increase in the crystallite size of the high metal loaded zeolites may be due to agglomeration of metal on the mouth of the pore or outside during reduction (ref.22).

Aben et al. (ref.23) reported from their studies on supported metal catalysts that the activity per exposed metal was independent of the metal crystallites or the support used. The observation on ruthenium is similar to that of Coenen et al. (ref.24) who noticed that the specific activity increased with decreasing crystallite size, except in the case of small crystallites where there is a possibility of metal support interaction.

XPS spectra

In Table 3, XPS parameters on 3-NiM are given. The unreduced NiM has a primary peak at 856.6 ev with FWHM of 4.4 ev. On reduction the peak has shifted to 857.1 ev with the indication of another one at 854.1 ev and the FWHM has broadened to 6 ev. The sattelite to primary line separation and the sattelite to

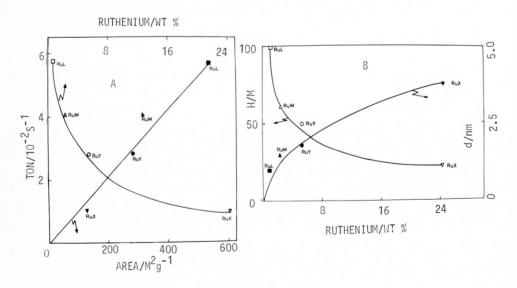

Fig. 2A   Variation of hydrogenation activity with metal area and metal content, at 353 K; LHSV = 2; H/HC = 2.4

2B   Variation of dispersion and crystallite size with ruthenium content primary intensity ratio have not changed after reduction. But the sattelite to primary width ratio has reduced. XRD revealed the presence of bulk Ni^0. However, the characteristic peak for Ni^0 at 852.8 ev with FWHM of 2 ev was not observed in the XPS spectra. Comparison of the binding energy values of 3-NiM with the literature values (ref.25-27) for $Ni2p_{3/2}$ suggests that there are more than one species of nickel viz. Ni^{2+} and possibly Ni^+ or $Ni(OH)_2$.

TABLE 3

XPS Parameters on nickel mordenite, 3-NiM

Catalyst	Binding energies, ev					ΔE_{sat}	Y_i	Y_w
	$Ni2p_{3/2}$+sattelite		C1s	Si2p	O1s			
UNREDUCED	856.6 (4.4)	862.1	285 (2.2)	103.7 (2.4)	532.5 (2.4)	6.0	0.35	0.96
REDUCED	857.1 (6.0)	862.8	285 (2.4)	103.3 (2.4)	532.3 (2.4)	5.7	0.31	0.68
	854.1 (2.9)							

Values in parantheses are FWHM (ev). ΔE_{sat}= satellite to primary line separation. Y_i and Y_w = satellite to primary ratios for intensity and width respectively.

ESR spectra

ESR spectra were taken at room temperature on unreduced and reduced 3-NiM after evacuation at 723 K for 2h. The g values for fresh and reduced samples were 2.22 and 2.42 respectively indicating Ni^{2+} state. The absence of any resolved parallel and perpendicular feature may be due to aggregation of Ni^{2+} ions to clusters on the surface. However, the asymmetry of the signal shows the presence of small percentage of Ni^+ which is expected to have a higher g value (g = 2.42).

CONCLUSIONS

1. Hydrogen adsorption can be used for measuring metal dispersion and crystallite size provided one is careful in choosing the monolayer volume of the adsorbate especially when the attainment of equilibrium is slow.

2. Ruthenium is rather easily reduced compared to nickel and gives a better dispersion. Calcination in oxygen prior to reduction decreases the dispersion of ruthenium.

3. Benzene hydrogenation activity depends on the available ruthenium area and

is independent of the zeolite type. In nickel-zeolite, the support seems to influence the catalytic activity.

4. Pretreatment of nickel zeolite in oxygen followed by hydrogen results in the formation of Ni^{2+} and Ni^+ or probably $Ni(OH)_2$ species.

ACKNOWLEDGEMENT

The author thanks Dr G Thyagarajan, Director, Regional Research Laboratory, Hyderabad for his encouragement and support and Mr K S Rama Rao for his assistance in adsorption measurements.

REFERENCES

1 M.A. Vannice, J. Catal., 44 (1976) 152-162.
2 S. Bhatia, N.N. Bakhshi and J.F. Mathews, Can. J. Chem. Eng., 56 (1978) 575-581.
3 H.H. Nijs, P.A. Jacobs, J.J. Verdonck and J.B. Uytterhoeven, Proc. 5th Int. Conf. Zeolites, Ed. L.V.C. Rees, (Heyden), 1980, 633-639.
4 R.J. Farrauto, AIChE Symp. Ser., 70 (1974) 9-22.
5 R.A. Dalla Betta, J. Catal., 34 (1974) 57-60.
6 J.L. Carter, J.A. Cusumano and J.H. Sinfelt, J. Phys. Chem., 70 (1966) 2257-2263.
7 D.J.C. Yates, W.F. Taylor and J.H. Sinfelt, J. Am. Chem. Soc., 86 (1964) 2996-3001.
8 K.C. Taylor, J. Catal., 38 (1975) 299-306.
9 Y.L. Lam and J.H. Sinfelt, J. Catal., 42 (1976) 319-322.
10 C.H. Bartholomew and R.J. Farrauto, J. Catal., 45 (1976) 41-53.
11 C.S. Brooks and G.L.M. Christopher, J. Catal., 10 (1968) 211-223.
12 B. Coughlan, S. Narayanan, W.A. McCann and W.M. Carroll, J. Catal., 49 (1977) 97-108.
13 H.P. Klug and L.E. Alaxander, 'X-ray Diffraction Procedures', Wiley, NY 1974.
14 D.G. Mustard and C.H. Bartholomew, J. Catal., 67 (1981) 186-206.
15 J.S. Smith, P.A. Thrower and M.A. Vannice, J. Catal., 68 (1981) 270-285.
16 J.T. Richardson, J. Catal., 21 (1971) 122-129.
17 P.J.R. Chutoransky and W.L. Kranich, J. Catal., 21 (1971) 1-11.
18 V. Penchev, N. Davidova, V. Kanazirev, H. Minchev and Y. Neinska, Adv. Chem. Ser., 121 (1973) 461-468.
19 M.F. Guilleux, D. Delafosse, G.A. Martin, and J.A. Dalmon, J.C.S. Faraday I, 75 (1979) 165-171.
20 N. Davidova, N. Peshev and D. Shopov, J. Catal., 58 (1979) 198-205.
21 D.L. Elliott and J.H. Lunsford, J. Catal., 57 (1976) 11-26.
22 Kh. M. Minachev, G.V. Antoshin, E.S. Shpiro and Yu.A. Yusifov, Proc. 6th Int. Congr. Catalysis (London) 2 (1977) 621-632.
23 P.C. Aben, J.C. Platteeuw and B. Stouthamer, Proc. 4th Int. Congr. Catalysis (Moscow) 1968 paper 31.
24 J.W.E. Coenen, R.Z.C. Van Meerten and H.T. Rijnten, Proc. 5th Int. Congr. Catalysis (Florida) 1 (1973) 671-681.
25 G. Ertl, R. Hierl, H. Knoezinger, M. Thiele and H.P. Urbach, Appl. Surf. Sci., 5 (1980) 49-64.
26 K.T. Ng and D.M. Hercules, J. Phys. Chem., 80 (1976) 2094-2102.
27 J.C. Vedrine, G. Hollinger and T.M. Duc, J. Phys. Chem., 82 (1978) 1515-1520.

EFFECT OF THE REACTION MEDIUM ON THE METAL MICROSTRUCTURE OF NICKEL-ZEOLITE CATALYSTS

N.P. DAVIDOVA, M.L. VALCHEVA and D.M. SHOPOV
Institute of Organic Chemistry, Bulgarian Academy of Sciences,
Sofia 1040 (Bulgaria)

ABSTRACT

On the basis of X-ray, IR spectral and derivatographic analysis data the formation and the state of the metal phase of nickel-zeolite catalysts are studied as a function of the composition, preparation mode and reaction medium. The catalyst samples are examined after a preliminary reduction treatment and after the reactions of carbon monoxide methanation, toluene disproportionation and diethyl sulphide hydrogenolysis took place. It is found that part of the reagents and/or reaction products indirectly (via an effect on the state of the zeolite lattice), other directly affect the state of the metal phase.

INTRODUCTION

It is known that the formation of nickel phase in nickel-containing catalysts in the presence of hydrogen yields the most regular crystallites (ref.1). Hence, the deviation in the microstructure of the latter should be related to other factors, e.g. carrier, way of production, additions, reaction medium etc.

In the present paper the effect of some of the mentioned factors is studied using a series of nickel-containing catalysts based on zeolites of type X, Y and mordenite. On the basis of X-ray data some crystallographic characteristics of the zeolite and of the metal phase are determined after preliminary reduction of the samples, after performing the reactions: toluene disproportionation, interaction between CO and H_2 and hydrogenolysis of diethyl sulphide and in some cases after subjecting to the effect of separate components of the reaction medium under the same conditions.

METHODS

The catalytic samples are obtained on the basis of synthetic zeolites of the type X, Y and mordenite with molar ratio

SiO_2/Al_2O_3 = 2.5, 5.0 and 10.0. The Ca-form of the zeolites (CaX and CaY) is obtained by ion exchange with Ca^{2+} from a 4N solution of $CaCl_2$. The extent of ion exchange is about 80 %. The Ni-form (NiCaX and NiCaY) is prepared by a further ion exchange of the Ca-form with Ni^{2+} from a 0.1N solution of $Ni(NO_3)_2$. The Ni-form of the mordenite (NiM) is prepared by ion exchange with Ni^{2+} from a 0.1N solution of $Ni(NO_3)_2$. The samples NiO/CaX, NiO/CaY and NiO/M are obtained by deposition of NiO on CaX, CaY and mordenite. The nickel content is kept nearly constant (about 3 wt %) in all samples.

X-ray analysis is performed with U = 40 kV and CuK_α irradiation. Infrared spectra of KBr-tabletted samples are taken in the region 400 - 1200 cm^{-1}. The derivatographic analysis is performed in the temperature range 20 - 1000°C with rate of 10°/min. The samples are prepared by cooling and passivating with argon after reduction (treating with hydrogen at 723 K for 2 h after the temperature is slowly elevated to this value) and after reaching constant activity towards the reactions: toluene disproportionation (temperature 703 K, space velocity 1.1 h^{-1} and H_2:toluene ratio of 10), interaction between CO and H_2 (temperature 703°K, space velocity 230 h^{-1} and CO:H_2 ratio of 1/3) and hydrogenolysis of diethyl sulphide (temperature 548 K, space velocity 3600 h^{-1} and flow rate of hydrogen 60 ml/min passing through a diethyl sulphide saturator kept at 0°C).

From the X-ray data following parameters are determined: zeolite unit cell dimension (a,b and c), nickel unit cell dimension (a), average size of the nickel particles (L) and intensity of the characteristic signal for (111)Ni (I). The deformation caused by the different size of the nickel particles (D_L) and the deformation due to the microtensions are calculated by the methods described in refs 2 and 3. Changes in the amount of Ni^o on the surface, in the form of particles sized $\geqq$ 5.0 nm were estimated from the relative change in the intensity of the characteristic signal for (111)Ni . As a measure of the deformation due to microtensions in the crystal (D_M) $\frac{\Delta \ell}{\ell}$ is considered, where l is the sum of the crystallographic face index (h, k and l) squares. As a measure of the deformation due to particle size (D_L) the normal logarithm of the intensity of mean sized nickel particles with no microtensions is used.

RESULTS AND DISCUSSION

The contact of the examined samples with hydrogen, then with the reaction medium leads to changes both in the zeolite framework and of the metal. From the values of the zeolite unit cell Parameter (Table 1) it is seen that after reduction with hydrogen for the zeolites type X and Y, the parameter "a" increases for the ion-exchanged samples due to metal phase formation also in the intracrystalline surface of the zeolite. For mordenite the parameter "c" is not altered and the significant changes in "a" and "b" are attributed to re-orientation of the system of narrow channels in the zeolite framework parallel to one of the latter parameters.

TABLE 1
State of the zeolite lattice.
Parameters of the unit cell (a, b and c) in air dry state (1), in reduced state (2) and after the reactions of carbon monoxide (3), toluene disproportionation (4) and diethyl sulphide hydrogenolysis (5).

Sample		Parameters of the unit cell (nm) in the state				
		1	2	3	4	5
NiCaX	a	2.482	2.494	2.472	2.498	2.475
NiO/CaX	a	2.481	2.485	2.470	2.485	2.474
NiCaY	a	2.466	2.478	2.456	2.479	2.463
NiO/CaY	a	2.464	2.463	2.452	2.464	2.458
NiNaM	a	1.780	1.666	1.850		
	b	2.016	1.864	2.029		
	c	0.750	0.759	0.759		
NiO/NaM	a	1.841	1.780	1.811		
	b	2.063	2.029	2.026		
	c	0.749	0.752	0.750		

The IR characteristic bands of the zeolite framework in the spectra of the samples are, however, not shifted, which signifies that the changes in the unit cell parameters caused by the metallic nickel, are due to a deformation without disturbing the structure. For the reactions of $CO + H_2$ interaction and hydrogenolysis of diethyl sulphide, compression of the zeolite framework is observed. This effect could be undoubtedly related to the action of water steam and H_2S, which are reaction products, since the same change in "a" is observed when they are separately used under the reaction conditions. Furthermore, in the IR spectra of the samples taken after the interaction between CO and H_2 and after the action of water steam under the same conditions, the bands of the asymmetrical valence vibrations of T-O at approx. 1000 cm^{-1} are shifted with about 40 cm^{-1} towards the higher wave numbers (for mordenite the shift is less). The DTA data for the molar specific heat of water evaporation from the zeolite structure of the samples indicate a decrease in zeolite hydrophility on elevating the SiO_2/Al_2O_3 ratio. Apparently, the destructive action of water steam is related to the zeolite hydrophility. The IR spectral data for samples subjected to the reaction hydrogenolysis of diethylsulphide, point, however, to a certain stabilization of the lattice - the bands are shifted towards the lower wave numbers. Simultaneously with these structural changes in the zeolite, or as a result of them, the state of the metal phase is altered.

It is known that the habit of the low-disperse nickel differs from that of high-disperse one (ref.4). There are data, that after mechanical mixing of NiO with zeolite and reduction only 1/4 of the nickel phase is present in low-disperse state (ref.5). The signal for (111) Ni of the samples NiO/CaX, NiO/CaY and NiO/M is considerably weaker compared to mechanical mixtures of NiO and CaX, CaY or M prepared by us, which signifies that the amount of low-disperse nickel in deposited samples is lower. It follows from Fig. 1, that in ion exchanged samples this amount is even less. It strongly depends both on the effect of the type and properties of the zeolite framework on the reducibility of Ni^{2+} and on the effect of the system of internal channels on the migration of Ni^{0}. There are sharp changes in the state of the metal phase under the action of the reaction medium. From the three model reactions (see Fig. 2), the conversion of toluene leads to the highest formation of low-disperse nickel. A relationship between the amount of low-disperse nickel and the selectivity towards the reaction of dealkylation is observed. For NiO/CaX and NiO/CaY, on which only dealkylation takes place, the intensity for (111)Ni is 1.8, resp. 2.6 times higher than for NiCaX and NiCaY, on which occurs preferably disproportionation (ref.6). The observed increase is less for the other two reactions. The difference in the habit of the low- and high-disperse nickel permits to use the relative change in signal intensity of (111)Ni as a measure for the relative changes in the part of the separate crystal faces in the total metal surface (ref.7). These changes raise changes in the catalytic behaviour of the metal phase towards reactions taking place preferably on definite crystallographic faces, e.g. the interaction between CO and H_2 is favoured by the face (111)Ni (ref.8).

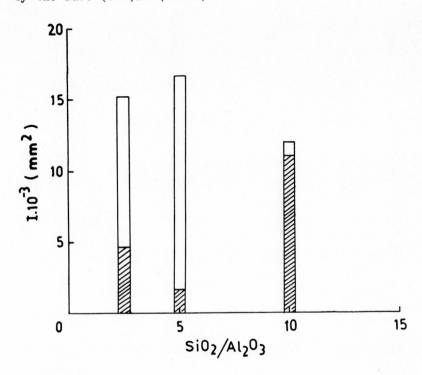

Fig. 1. Effect of the molar ratio SiO_2/Al_2O_3 of the zeolites on the amount of low-disperse nickel /I, (111)Ni/ for the samples obtained by ion exchange ▨ and by deposition ☐ in reduced state.

In Table 2 are presented some characteristics of the low-disperse nickel, which reflect the metal microstructure after reduction and after the action of the reaction medium. In the examined case the size of the particles is below 100 nm and therefore they are subjected to deformations due to microtensions (D_M), which affect the internal crystal energy and to deformations caused by the particles size (D_L). When the preparation mode is kept constant, the microtensions increase on rising the ratio

SiO_2/Al_2O_3. For constant molar ratio the microtensions are higher for the ion exchanged samples. The deformation due to particle size depends analogously on the same factors, which cause changes in L.

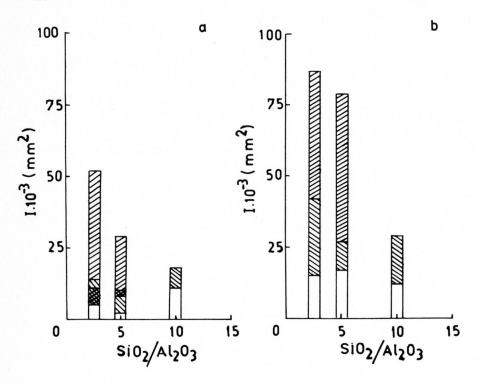

Fig. 2. Effect of the molar ratio SiO_2/Al_2O_3 of the zeolites on the amount of low-disperse nickel /I, (111)Ni/ for the samples obtained by ion exchange (a) and by deposition (b)

   □   in reduced state
   ▨   after the methanation of carbon monoxide
   ▨   after the disproportionation of toluene
   ✳   after the hydrogenolysis of diethyl sulphide

In order to elucidate the observed differences in the metal microstructure for the three model reactions, the effect of the separate components of the reaction medium is studied. For instance, in the interaction between CO and H_2 the water steam and CO lead to compression of the zeolite lattice, deceleration of the agglomeration of the metallic nickel, decrease in micro-tensions and for the ion exchanged samples to the appearance of

new amounts of low-disperse nickel on the zeolite surface. Methane
has the reverse effect - increase in the mean size of nickel
particles, increase in the microtensions in them and increase in
the amount of low-disperse nickel phase.

TABLE 2

State of the metal phase.

Parameter of the unit cell (a), size of the metal particles (L),
deformation due to the microtensions (D_M) and deformation caused
by the different size of the metal particles (D_L) in reduced
state (1) and after the reactions of carbon monoxide methanation
(2), toluene disproportionation (3) and diethyl sulphide hydrogenolysis (4) .

Sample	State	a, nm	L, nm	D_M	D_L
NiCaX	1	0.3528	18	0.492	7.28
	2	0.3518	24	0.468	8.36
	3	0.3527	50	0.507	9.70
	4	0.3522	25	0.480	8.15
NiO/CaX	1	0.3522	84	0.480	7.30
	2	0.3530	74	0.482	10.60
	3	0.3522	78	0.480	10.69
NiCaY	1	0.3518	20	0.530	6.70
	2	0.3518	18	0.490	6.45
	3	0.3518	35	0.490	6.92
	4	0.3519	20	0.482	6.47
NiO/CaY	1	0.3534	42	0.491	9.00
	2	0.3519	36	0.482	8.50
	3	0.3534	60	0.492	9.42
NiNaM	1	0.3526	25	0.532	8.65
	2	0.3524	15	0.593	7.75
NiO/NaM	1	0.3524	70	0.502	8.30
	2	0.3526	63	0.512	7.44

Since the changes in the zeolite structure and in the microstructure of the low-disperse nickel phase have the same direction
for the interaction between CO and H_2, as in the case of the
effect of the water steam, they could be considered as determining the state of the whole catalyst. In the hydrogenolysis of
diethyl sulphide most important is the effect of H_2S, which leads
to a decrease in the parameter of the zeolite unit cell, decrease
in the mean size of nickel particles and significant decrease of
the microtensions in the metal. For the conversion of toluene

the changes in the metal state are determined by the effect of the methane.

Hence, it could be concluded that the reagents affecting simultaneously the zeolite framework and the metal phase determine the metal microstructure during the reaction course, whereas the reagents affecting only the metal phase, are significant only for the re-distribution of the low-disperse nickel phase.

REFERENCES

1 G.D. Zakumbaeva, Interaction of Organic Compounds with the surface of Group VIII Metals, Nauka, Alma-Ata, 1978, 26 pp.
2 G.B. Bokij and M.A. Poray-Koshiz, X-ray Analysis, Nauka, Moskow, vol. 1, 1974, 441 pp.
3 S.S. Gorelik, L.N. Rastorguev and Y.A. Skakov, X-ray and Electronograph Analysis, Metallurgia, Moskow, 1970, 83 pp.
4 R. Van Hardeveld and A. van Montfoort, Surface Sci., 4 (1966) 396.
5 P. Low and C.N. Kenney, J. Catal., 64 (1980) 241.
6 N.P. Davidova, N.V. Peshev and D.M. Shopov, J. Catal., 58 (1979) 198.
7 N.P. Davidova, M.L. Valtcheva and D.M. Shopov, Zeolites, 1 (1981) 72.
8 J.A. Dalmon and G.A. Martin, J. Chem. Soc. Faraday Trans. I, 75 (1979) 1011.

ACTIVITIES AND SELECTIVITIES OF REDUCED NiCaX FAUJASITES IN THE CARBON MONOXIDE HYDROGENATION REACTION

H. SCHRÜBBERS, G. SCHULZ-EKLOFF and G. WILDEBOER
Research Group Applied Catalysis, University of Bremen,
D-2800 Bremen 33

ABSTRACT

The activities (mg hydrocarbons/g nickel·h) and selectivities (number of hydrocarbons) of a series of nickel plus calcium exchanged and reduced faujasites in the carbon monoxide hydrogenation reaction have been determined. A close correlation between nickel ions in SI sites prior to the reduction and the initial activities is found and referred to a preferential reducibility of nickel in the hexagonal prisms. The observed temperature dependence of the selectivities is indicative for a lower structure-sensitivity of the chain growth via carbon monoxide insertion as compared to the methanation.

INTRODUCTION

Nickel zeolites have been found to be active catalysts in the carbon monoxide hydrogenation reaction (ref. 1,2). It has been elaborated that the extent of reduction of the nickel component depends on many parameters, like e.g. ion exchange procedures, dehydration conditions, type and distribution of additional cations, etc. (ref. 2,3,4). In the following the influence of additional calcium ions on the activities and selectivities of reduced NiCaX faujasites is reported and discussed.

EXPERIMENTAL
Catalysts

The NaX zeolite was prepared by hydrothermal synthesis. Ion exchange was carried out in solutions of $Ca(CH_3COO)_2$ and $Ni(CH_3COO)_2$. The extent of ion exchange was determined by atomic absorption spectroscopy (see Table 1). Commercial Ni/Al_2O_3 (Strem Chemicals, Newbury Port) was studied by comparison.

TABLE 1

Composition of the catalysts

Abbreviation	Composition	wt%Ni	Remarks
NiCaX5	$Ni_{9.6}Ca_{10.6}Na_{45.5}X$	4.2	Simultaneous exchange of Ca and Ni (1x)
NiCaX1.1	$Ni_{10.1}Ca_{10.7}Na_{45.5}X$	4.4	Consecutive exchange of Ca(1x) and Ni(1x)
NiCaX3.1	$Ni_{8.6}Ca_{20.8}Na_{27.1}X$	3.8	Consecutive exchange of Ca(3x) and Ni(1x)
NiCaX1.2	$Ni_{20.3}Ca_{8.9}Na_{27.6}X$	8.7	Consecutive exchange of Ca(1x) and Ni(2x)
NiCaX10	$Ni_{21.3}Ca_{20.4}Na_{2.6}X$	9.2	Ca exchange, activation, Ni exchange
Ni/Al_2O_3	-	11.8	-

Activation, reactor and analysis

The catalysts were activated in argon ($420^{o}C$, 16 h, heating rate: $5^{o}C/min.$) and reduced with hydrogen ($300^{o}C$, 25 h, 1 bar). The reaction was carried out in a fluidized bed reactor (1 bar, CO/H_2 = 3/7, GHSV 300) at 250, 300 and $350^{o}C$. The reaction products were analyzed by capillary gas chromatography.

Characterization

The samples and catalysts were characterized by thermal analysis, magnetic susceptibility, X-ray analysis (ref. 2) and transmission electron microscopy (ref. 5,6).

RESULTS

Nickel ion distribution and catalytic activity

The fraction of nickel ions in SI sites of the faujasite lattice prior to the reduction has been estimated from the measured magnetic moments (ref. 2), and is correlated with the yield of hydrocarbons (mg hydrocarbons/g nickel·h), which will be obtained in the carbon monoxide hydrogenation at 250°C and 1 h time on stream immediately after the reduction step, and which is defined as "initial activity" (see Fig. 1). The obtained rather strict correlation is surprising in view of the many parameters affecting the reduction process.

The quotient of the yields at 300°C and 250°C gives the values 3.1 (Ni/Al$_2$O$_3$), 6.9 (NiCaX1.2), 7.9(NiCaX5), 8.2 (NiCaX10), 23.6 (NiCaX3.1) and 25.8 (NiCaX1.1). From the fact that this ratio increases with decreasing initial activity a close connection between initial activity and degree of reduction, which is always incomplete for nickel faujasites (ref. 3), can be concluded. A direct determination of the degree of the reduction via Ni^0 oxidation by dichromate failed due to large errors.

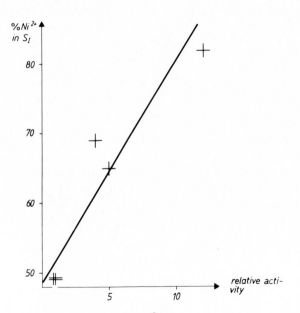

Fig.1. Fraction of Ni^{2+} in SI sites prior to reduction (ref. 2,6) for NiCaX1.1 (49%), NiCaX3.1 (49%), NiCaX5 (65%), NiCaX1.2 (69%) and NiCaX10 (82%) correlated to the relative initial activities in the carbon monoxide hydrogenation (1 bar, CO/H$_2$ = 3/7, GHSV 300).

Temperature dependence of the selectivities

Fig. 2 gives the number of hydrocarbons formed in the carbon monoxide hydrogenation reaction at 250, 300 and 350°C. Differences in the number of hydrocarbons are mainly due to the differences in the number of carbon atoms in the built up hydrocarbons and do not refer to differences in the kind of isomers from molecules with a given number of carbon atoms. At 250°C the studied catalyst can be classified in one group with a marked absolute amount of C_{6+} products, exhibiting a chain growth probability $\alpha \approx 0.4$ (Ni/Al_2O_3; NiCaX5), and in another group with C_1 - C_5 molecules, a negligible absolute amount of C_{6+} hydrocarbons, and a value of $\alpha = 0.25 - 0.3$.

At 250°C the most active catalysts produce the highest number of hydrocarbons, whereas at 350°C the reverse effect has to be stated.

DISCUSSION

The correlation between the fraction of nickel ions in SI sites prior to the reduction and the initial activity in the carbon monoxide hydrogenation (see Fig. 1) supports previous reports, which assume a favored reducibility of nickel ions in SI sites (ref. 2, 4,7).

The temperature dependence of the selectivities has to be discussed on the basis of the mechanisms, which have been proposed for the carbon monoxide hydrogenation on nickel (ref. 8 and cited lit.).

Carbon monoxide can be either dissociated to form methane preferentially or can be inserted in a H_xC-Ni bond to give C_{2+} hydrocarbons (ref. 9). The observed temperature dependence of the selectivities can be understood, if it is assumed that a rapid carbon monoxide dissociation at temperatures above 600 K (ref. 10) for a subsequent methanation decreases the probability for a carbon monoxide insertion. An alternative interpretation has to take into account a high hydrogenolysis activity at the centers of high methanation activity resulting in a rapid hydrocracking of initially formed C_{2+} hydrocarbons. Both models require different centers for the carbon monoxide dissociation respectively the hydrocracking, on one hand, and the insertion, on the other hand. Since surface nickel atom ensembles, which are needed for structure-sensitive reactions (ref. 11) are assumed to be responsible for the former reactions (ref. 9,10), a low structure-sensitivity might be postulated for the carbon monoxide insertion. Support for this postulation can be found in the observation, that strong sintering (NiCaX3.1, NiCaX1.2, NiCaX1.1) and coke deposition (NiCaX10, used in the Fischer-Tropsch-reaction, ref. 12) increase the number of C_{2+} products at $350^{\circ}C$ (see Fig. 2; ref. 6). Both inhibiting processes might preferentially decrease the centers for the methanation and the hydro-cracking, because these centers have to catalyze dissociations of strong bonds and should therefore exhibit largely unsaturated and highly reactive surface metal atoms respectively ensembles.

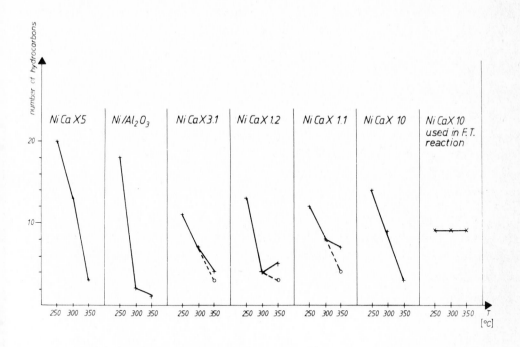

Fig. 2. Temperature dependence of the number of hydrocarbons formed in the carbon monoxide hydrogenation (1 bar, CO/H_2 = 3/7, GHSV 300). At $350^{\circ}C$ the number increases with the time on stream from 1h(o) to 6h(+) on those catalysts (NiCaX3.1, NiCaX1.2, NiCaX1.1), which exhibit a corresponding increase in the mean particle diameter by sintering (ref. 6).

ACKNOWLEDGMENT

The authors wish to thank R. Nowak for the magnetic susceptibility measurements and many helpful discussions. A critical revision by Prof. N. Jaeger is gratefully acknowledged.

REFERENCES

1   D.J. Elliott and J.H. Lunsford, J. Catal. 57 (1979), 11
2   N. Jaeger, U. Melville, R. Nowak, H. Schrübbers and
        G. Schulz-Ekloff, Stud. Surf. Sci. Catal. 5 (1980), 335
3   M. Briend-Faure, M.F. Guilleux, J. Jeanjean, D. Delafosse,
        G.D. Mariadassou and M. Bureau-Tardy, Acta Phys. et Chem.,
        Szeged 24 (1978), 99
4   C. Minchev, V. Kanacirev, L. Kosova, V. Penchev, W. Gunsser
        and F. Schmidt, in: Proc. 5th Int. Conf. on Zeolites, (Ed.:
        L.V. Rees), Heyden, London 1980, 255 - 363
5   D. Exner, N. Jaeger, R. Nowak, H. Schrübbers and G. Schulz-
        Ekloff, J. Catal. 74 (1982), 188
6   H. Schrübbers, Thesis, Universität Bremen 1982
7   T.A. Egerton and J.C. Vickerman, J.C.S. Faraday Trans. I,
        69 (1973), 39
8   P. Biloen and W.M.H. Sachtler, Adv. Catal. 30 (1981), 165
9   V. Ponec, Catal. Rev.-Sci. Eng., 18 (1978), 151
10   J.A. Dalmon and G.A. Martin, J.C.S. Faraday Trans. I, 75
        (1979), 1011
11   M. Boudart, Adv. Catal., 20 (1969), 153
12   H. Schrübbers, P.J. Plath und G. Schulz-Ekloff, React.
        Kinet. Catal. Lett., 13 (1980), 255

FISCHER-TROPSCH SYNTHESIS ON POLYFUNCTIONAL MANGANESE/IRON - PENTASIL ZEOLITE CATALYSTS

K. MÜLLER[1], W.-D. DECKWER[2] and M. RALEK[1]

[1]Institut für Technische Chemie, Technische Universität Berlin, Straße des 17. Juni 135, D 1000 Berlin 12 (Germany)

[2]Fachbereich Chemie, Universität Oldenburg, P.O. Box 2305, D 2900 Oldenburg (FR Germany)

ABSTRACT

Owing to the high contribution of syngas costs to the economy of all indirect liquefaction routes a process is desired which could directly use syngas from second-generation gasifiers with low hydrogen content to produce a C_5 to C_{11} hydrocarbon mixture, which can be used as a high RON fuel and as a feedstock for chemicals. The Fischer-Tropsch synthesis on the polyfunctional MnFe-ZSM-5 catalysts yields products containing about 20 % wt. structure isomers and about 25% wt. aromates. The reaction follows the nontrivial polystep mechanisms.

INTRODUCTION

In a recent study of Poutsma (1) a new approach has been proposed to directly produce hydrocarbon mixtures rich in aromates from synthesis gas. These hydrocarbons can be used either as a high RON motor fuel or as a feedstock for chemicals. With regard to the above proposal the primary synthesis products from Fischer-Tropsch (FT) catalyst are converted on zeolites of pentasil type to structure isomers and aromates. In the two stage process and a single stage process concept as well the FT catalyst has to fulfill certain requirements. As an essential presupposition, the Fischer-Tropsch catalyst should be operated under conditions to give the lightest, most olefin and oxygenate rich product possible and at the same time provide for a low aging rate and a low coking tendency. Such a catalyst must produce these reactive species from CO-rich synthesis gases of second generation coal gasifiers, that means, it must retain high shift activity in order to keep the CO/H_2 feed ratio comparable to the usage ratio.

In the two stage process concept the output from the Fischer-Tropsch reactor is used as feed for the Mobil-Gasolin process (2), hence each reactor can be operated at its optimum conditions. Therefore, such a two-step process has the definite advantage that each process step can be carried out at different and individually optimized conditions. In addition, catalyst regeneration can be done at the specific conditions required for each catalyst. In a single stage process, both catalysts are placed in the same reactor. Therefore, synergistic interactions between the different catalyst functions occur and this in turn may lead to better activity and improved selectivity. However, a single stage process does not permit to adjust the operating conditions to the individual reaction conditions. This may involve serious problems. For instance, for converting methanol on ZSM-Zeolite, temperatures of 370 to 400°C are reported in the literature (2) while the FT Synthesis on conventional iron and cobalt catalysts is performed at temperatures of 200 to 270°C. It is therefore desirable that in a combined single stage process the FT catalyst has a sufficient activity and lifetime at higher temperatures, e.g. above 300°C at least.

In order to meet all these particular requirements, it is believed that the selective manganese-iron precipitation catalysts developed by Kölbel et al. (3) are more appropriate than fused iron catalysts (4) and iron impregnated zeolitic catalysts (5). These precipitated Mn/Fe-catalysts approach their entire synthesis activity at temperatures beyond 290°C and in addition they have the ability to convert CO rich syngas (CO/H$_2$<1.7) to hydrocarbons with a sufficient lifetime in fixed bed reactors as well (6). The hydrogenation activity is reduced, therefore the produced hydrocarbons have an olefin fraction of up to 70 % (7, 8). The synthesis carried out at about 290°C gives a product slate which follows Schulz-Flory distribution with a chain growth probability of α = 0.67 (9) to 0.46 (with partial sulfur covering (10)). From magneto-chemical measurements, ESR and Mößbauer spectroscopy one can conclude that the Mn/Fe catalyst consists of fine distributed iron compounds (ε-Fe$_2$C, Fe$_5$C$_2$/FeC, Fe$_3$O$_4$) embedded in a MnO matrix (11). XPS and SIMS investigations have shown that the concentration of the iron phases is significantly lower at the surface than in the catalyst bulk. The iron depletion at the surface causes the formation of isolated clusters (12). This phenomenon and the strong metal support interaction (SMSI) are likely responsible for the catalytic properties of the Mn/Fe catalysts. These considerations present the starting point of the investigations presented here, e.g., a study on the performance of the Mn/Fe catalyst in combination with zeolites of pentasil type in the single and two stage process well. The catalytic properties of these zeolites used are described in detail in the literature (13, 14).

EXPERIMENTAL

The Mn/Fe catalyst has been prepared by the continuous precipitation technique (15) and optimized by Lehmann (8). The precipitate is a mixture of γ-Mn_2O_3 and α-Fe_2O_3. The iron content was 9 % wt. The zeolites ZSM-5 (Si/Al = 40) and silicalite (Si/Al>800) were prepared by Maiwald (16). Calcination of both zeolites was done at $550^{\circ}C$ in oxygen atmosphere for 24 h. The sodium ions were exchanged with protons by treatment with 1 nHCl. By using a ball mill the size of the various catalytic components was reduced to 0.005 to 0.01 cm. For studying the catalytic properties in the fixed bed the small particles were pressed to pellets either for each catalytic component separately or for a physical mixture of both components, e.g. Mn/Fe catalyst precursor and ZSM-5 or silicalite, respectively. For obtaining a more intimate mixing and penetration of the catalytic species, the Mn/Fe catalyst was precipitated in the presence of ZSM-5. In addition, pellets have been prepared from larger particles with diameters of 0.05 to 0.075 cm in order to detect possible internal mass transfer limitations.

In case of the single stage process variant 40 cm^3 of catalyst mixture were used, the volume fractions of the zeolite and the FT catalysts being equal. In the two stage arrangement 20 cm^3 of the zeolite were firstly introduced into the reaction tube and after a layer of glas wool 20 cm^3 of the Mn/Fe catalyst was added The synthesis gas flows downwards, at first, through the bed with the FT catalyst and after that through the zeolite catalyst.

Catalyst activation and synthesis were done in the same reactor successively. Catalyst activation involved 3 steps: Under nitrogen flow the catalyst was heated up to $260^{\circ}C$ and calcinated for several hours. Carburation of the FT catalyst was achieved by formation with CO for 20 hours at $260^{\circ}C$, 1.8 bars and a space velocity of 800 h^{-1}. This was followed by reducing with hydrogen for 20 hours at the same temperature and pressure and a space velocity of 1600 h^{-1}.

TABLE 1

Results on Mn/Fe - Pentasil zeolite systems.

Catalyst Arrangement	Mn/Fe	Mn/Fe-silicalite		Mn/Fe-ZSM-5	
		single stage	two stage	single stage	two stage
Temperature °C	312	310	310	312	312
Pressure bar	12	12	12	12	12
Space velocity h⁻¹	450	340	340	350	360
Inlet ratio CO/H2	1.42	1.55	1.55	1.44	1.43
Conversion (CO + H2)	0.80	0.87	0.88	0.78	0.86

Yields in g/Nm³ syngas fed into the reactor

	Mn/Fe	Mn/Fe-silicalite		Mn/Fe-ZSM-5	
		single stage	two stage	single stage	two stage
All hydrocarbons	164.6	184.2	186.4	162.3	180.8
Methane	33.0	37.2	26.1	42.2	31.5
Ethylene	4.2	2.9	3.8	0.1	0.9
Propylene	23.3	5.2	5.7	0.2	2.5
Butylenes	11.6	11.2	11.5	0	5.3
ΣC2-C4 olefins	39.1	18.5	21.0	0.3	8.7
Ethane	16.0	20.7	15.5	10.1	15.9
Propane	7.3	8.5	6.9	13.4	8.1
Butenes	8.1	10.3	10.7	19.9	11.0
ΣC2-C4 paraffins	31.4	39.5	33.1	43.4	35.0
Liquid phase, C5+	61.0	89.0	106.2	76.4	105.7

Yield in liquid phase C5+, g/Nm³ syngas fed into the reactor

	Mn/Fe	Mn/Fe-silicalite		Mn/Fe-ZSM-5	
		single stage	two stage	single stage	two stage
Structure isomers	20.7	70.0	82.8	32.4	75.0
Nonbranched HC	40.3	14.0	23.4	8.9	20.1
Aromates	0	5.0	0	35.1	10.6
Olefins	37.8	53.3	63.7	0	45.4
Paraffins	23.2	30.7	42.5	41.3	49.7

The synthesis runs were carried out at 12 bars and a space velocity of about 350 h^{-1}. This value refers merely to the volume of the Mn/Fe catalyst in the bed as only this catalyst is responsible for the primary reactions of the synthesis gas.

RESULTS AND DISCUSSION

Synthesis conditions and results obtained after a run time of 200 h are summarized in table 1. For comparison the table give also the data achieved for Mn/Fe catalysts alone. In the two process modes the application of the silicalite leads predominantly to isomerisation of the primary FT products and to higher yields of the C_{5+} fraction. The products slates do not differ much for the single and two-stage processes. Small amounts of aromates are produced in the one-stage process variant. However, for the catalyst systems with the Al rich ZSM component noticable yields of aromates could be obtained, especially in the one-step process mode with 45 g of aromates per Nm^3 synthesis gas converted. If refered to the entire amount of hydrocarbons produced this corresponds to a carbon selectivity of 27 %. The detailed composition of the aromatic product fraction is: toluene (13%wt.), xylene (29%wt.), ethyltoluene (25%wt.), trimethylbenzene (11%wt.), other aromates (22%wt.). Similar results have been found with the coprecipitated Mn/Fe-ZSM-5 catalysts. It should be emphasized that the production of aromates is simultaneousely accompanied by a loss of the yield of branched hydrocarbons and a drastic decrease of olefin yields. Also the yield of isomers is remarkably reduced in the single-stage process compared with the two-stage process.

The remarkable differences of selectivity with regard to olefins, isomers, and aromates point out that the single-stage synthesis with the physical mixture of both catalysts runs via a non-trivial polystep reaction mechanism (4, 17). As to present days mechanistic views short chain olefins like propylene and butylenes play an important role in the formation of aromatic compounds (13, 18). For instance, short chain olefins are probably directly involved in a mechanismus leading to aromatic compounds as they were converted predominantly to aromates. In contrast, long chain hydrocarbons were isomerized and aromate formation occurs only to a less extent. With consideration of these phenomena it is possible to understand qualitativly the observations concerning the synthesis runs with the intimate mixture of both catalysts. Fig. 1 presents some reaction step selected out of a more comprehensive system of FT-reactions which take place simultaneously.

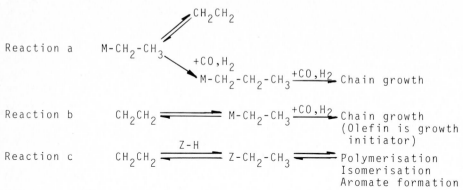

Fig. 1. Selected single steps of the polystep reaction mechanism (Z active site on the zeolite surface)

Due to the mixing of both catalytic components, olefins desorbing from the surface of the Mn/Fe catalyst can be readsorbed by the zeolite component and reaction c may occur. This, inturn, affects the olefin desorption equilibrium involved in reaction step a. Therefore, the concentration of the complexes adsorbed at the surface decreases. As all steps of the chain growth mechanism are affected by this equilibrium and the readsorption at the zeolite,the formation of longer chains by surface reaction will be hindered. Thus, only short chain olefins reach preferably the acidic sites of the zeolites where they are converted to aromates without much isomerisation. On the other hand, if the two catalyst are arranged in two zones one after the other the zeolitic component has no possibility to influence anyhow the reactions occuring at the surface of the Mn/Fe catalyst. In the first catalyst layer (Mn/Fe) larger amounts of alkanes and long chain olefins are already formed which can be converted to aromates in the ZSM-5 layer only to a small degree.

The arrangement of the individual catalytic components has a significant influence on the obtained product slate and one can suspect that the effective diffusion of the olefin intermediate within the catalyst particles plays also a dominant role. Therefore, additional measurements under variation of the particle size were carried out (19). Thus, the diffusion paths of the reactands,

intermediates, and products could be changed, under otherwise
identical conditions, the runs were performed with particle sizes
of 0.05 to 0.07 cm as compared to 0.005 to 0.01 cm used in the
above studies. The comparison of the product distributions obtained with the two particle sizes clearly demonstrates the considerable differences in selectivity of aromates, isomers, and
olefins. While the products of the small particle catalyst system
are free of olefins, an olefin selectivity of 50 % is found for
the large particles. One can conclude from this result that the
effective reaction rate of the olefin intermediates has been reduced by diffusional limitations in the larger catalyst particles.
For this catalyst, the desorption processes at the Mn/Fe surface
are less affected and this leads to an increased formation of
longer chain Fischer-Tropsch products.

The single-stage process variant with fine dispersed catalyst
particles is obviously characterized by higher yields of aromates.
However, its technical applicability depends on solving the task
of regenerating simultaneously both catalytic components - a problem which is encountered with all Fischer-Tropsch catalysts of
zeolite type.

Acknowledgement

Thanks are due to the Ministry of Research and Technology of
the Federal Republic Germany for financial support.

LITERATURE

1. M.L. Poutsma, Oak Ridge National Laboratory, Report No 5635
   (1980)
2. C.D. Chang, J.C.W. Kuo, W.H. Lang, S.M. Jacob, J.J. Wise,
   A.J. Silvestri, Ind.Eng.Chem., Process Des.Dev. 17, 225 (1978)
3. Belg.Pat. 837628 (H. Kölbel, K.-D. Tillmetz, 1975), H. Kölbel,
   M. Ralek, K.-D. Tillmetz, Proc. 13th Intersoc. Energy Conv.
   Conf., Soc.Automat.Eng., San Diego 1978, p. 482
4. P.D. Caesar, J.A. Brennan, W.E. Garwood, J. Ciric, J.Catalysis
   50, 274 (1979)
5. V.U.S. Rao, R.J. Gormley, Hydrocarbon Processing, November
   1980, p. 139
6. H.-J. Lehmann, Diplomarbeit, Institut für Technische Chemie,
   Technische Universität Berlin, 1975
7. G. Lohrengel, M.R. Dass, M. Baerns, Preparation of Catalysts,
   p. 41, Elsevier 1979; G. Lohrengel, Thesis, Ruhr-Universität
   Bochum,1980
8. H.-J. Lehmann, Thesis, Technische Universität Berlin, 1981
9. W.-D. Deckwer, H.-J. Lehmann, M. Ralek, B. Schmidt, Chem.-Ing.-
   Tech. 53, 813 (1981)
10. H.-J. Lehmann, H. Nguyen-Ngoc, M. Ralek, Chem.-Ing.-Tech. 54,
    52 (1982)

274

11. W. Podestà, Diplomarbeit, Institut für Physikalische Chemie, Universität Hamburg, 1980
12. W. Benecke, Thesis, Technische Universität Berlin, 1982
13. J.R. Anderson, K. Foger, T. Mole, R.A. Rajadhyaksha, J.V. Sanders, J. Catal. 58, 114 (1979)
14. C.J. Chang, A.J. Silvestri, J. Catal. 47, 249 (1977)
15. H. Kölbel, M. Ralek, Catalysis Rev.-Sci.Eng., 21 (2), 225 (1980)
16. W. Maiwald, Thesis, Universität Hamburg, 1980
17. P.B. Weisz, Adv. Catal. 13, 137 (1962)
18. E.G. Derouane, J.B. Nagy, P. Deyaifve, J.H.C. van Hooff, B.P. Spekman, J.C. Vedrine, C. Naccache, J. Catal. 53, 40 (1978)
19. K. Müller, Diplomarbeit, Institut für Technische Chemie, Technische Universität Berlin, 1981

OXIDATION OF ETHYLENE ON SILVER-LOADED NATURAL ZEOLITES

G. BAGNASCO[1], P. CIAMBELLI[2], E. CZARAN[3], J. PAPP[3], G. RUSSO[1]

[1]Institute of Industrial Chemistry and Chemical Plants, University of Naples,
P. Tecchio, 80125 Naples (Italy)

[2]Institute of Chemistry, University of Naples, v. Mezzocannone, 4, 80134 Naples
(Italy)

[3]Central Research Institute for Chemistry of the Hungarian Academy of Sciences,
P. O. Box 17, H-1025 Budapest (Hungary)

ABSTRACT

The structural characterization and catalytic activity of silver-loaded italian chabazite and hungarian mordenite in the oxidation of ethylene to ethylene
oxide are reported. The catalysts were characterized as pore volume by N_2 adsorption, silver crystallite size by oxygen chemisorption, X-ray diffraction line
broadening, electron microscopy. Catalytic performance at 240°C, 3% ethylene
concentration in air, was tested. Silver crystallite size ranged from some ten
to several hundred Angströms, depending on the type of zeolite and silver concentration. Silver zeolites were active and selective in the epoxidation of ethylene. Selectivity can be related to silver crystallite size.

INTRODUCTION

Selective oxidation of ethylene to ethylene oxide over silver catalysts is a
very important and well-established industrial process, but many fundamental questions concerning the reaction have not yet been answered(ref.1,2).

Industrial catalysts consist of silver deposited on a low-surface-area support.
High-surface-area supports were reported up to 1971 as unsatisfactory due to the
low selectivity exhibited by catalysts prepared by this way. Silver catalysts
supported on silicas with high surface area were investigated by Harriott (ref.3).
Some of these catalysts resulted more active and about as selective as silver on
low-surface-area supports. Low selectivity obtained on other medium or high-surface-area supports was attributed to the very small size of silver particles
(ref. 3).

Among the various porous high-surface-area supports zeolites are unique in the porous structure and this makes interesting their use in metal catalyst formulation. Recently Giordano et al.(ref.4,5) showed that selective silver catalysts can be prepared by impregnating commercial molecular sieves, while silver-exchanged zeolites produce only total oxidation.

In this paper structural characterization and catalytic performance of silver-loaded italian natural chabazite and hungarian natural mordenite are reported.

METHODS

Materials

The supports for catalysts preparation were a natural chabazite contained in a tuff deposit of Southern Italy (sample 1 in ref.6) and a natural mordenite from the Tokaj-mountains (Hungary)(ref.7). Non-zeolitic components in the rocks are mainly quartz, pyroxenes, amorphous glass.

Silver catalysts were prepared by impregnating samples of these zeolites(60-80 mesh size) with $AgNO_3$ aqueous solutions. The impregnated samples, after drying at 110°C for 15 h, were loaded to a glass reactor where they were heated in dry air at 350°C overnight and then reduced with 7% H_2 in N_2 at 350°C for 2 h, immediately before adsorptive or catalytic tests.

Physico-chemical characterization

The content of silver in the catalysts was analyzed by a modified Mohr method.

The porosity of the catalysts was characterized by N_2 adsorption at -196°C, using volumetric or flow chromatographic techniques. Catalyst samples were examined before and after reduction.

X-ray diffraction spectra were obtained by a Philips PW 1016 recording apparatus . Silver particle size was calculated by the half-peak diffraction line width corrected for the instrumental broadening.

The average crystallite size and size distribution were estimated from electron micrographs taken with a Zeiss EF or a Siemens Elmiscope 102 microscope.

The surface area of metallic silver and the average size of silver crystallites were obtained from oxygen chemisorption. The adsorption tests were conducted in a glass flow apparatus at 240°C and 10 torr O_2 partial pressure in He stream. Before O_2 chemisorption the catalysts were treated in dry air and then reduced as described above; purged at 350°C for 2 h with purified He, fed through an Oxysorb system to keep oxygen concentration lower than 0.1 ppm; cooled to 240°C always in He flow. The chemisorbed O_2 was calculated from its breaktrough curve, as obtained from gas-chromatographic analysis. The stoichiometry of the chemisorption was taken as one oxygen atom per silver surface atom (ref.8). Blank O_2 adsorption runs were carried out on the supports to correct chemisorption data.

Catalytic activity

The catalytic tests were carried out in a pyrex flow reactor equipped with a thermocouple well, immersed in a constant temperature bath. The catalyst was diluted with inert alumina in the ratio 1 to 10. The runs were conducted at 240°C, atmospheric pressure, feed composition of 3% ethylene in dry air. Gas-chromatographic analysis detected the concentration of CO_2, C_2H_4, C_2H_4O and acetaldehyde.

RESULTS AND DISCUSSION

Some properties of the catalysts are given in Table 1. Silver crystallite sizes confirm the tendency of metallic silver to migrate and agglomerate on the external surface of zeolite, as already found by Giordano (ref.4). This conclusion is supported by N_2 adsorption results. In Fig. 1 the adsorption isotherms of N_2 on two silver chabazite catalysts, obtained before and after H_2 reduction, are reported. Reduction results in silver migration out of the cavities of zeolites.

TABLE 1

Catalysts properties[a].

Catalyst	Silver (wt%)	S^b (m^2Ag/g cat.)	$d_0{}^b$ (A)
C-5	4.6	4.7	56
C-10	9.4	11.8	48
C-20	20.5	9.9	118
C-30	27.1	8.2	189
M-2	2.0	2.0	57
M-4	3.5	3.4	58
M-8	7.7	4.9	89

[a]Catalysts reduced as described in the text. C = Chabazite, M = Mordenite.
[b]From oxygen chemisorption.

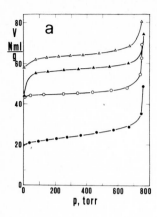 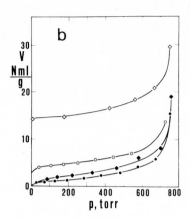

Fig. 1. Adsorption isotherms of N$_2$ (T=-196°C) on the catalysts before (full symbols) and after reduction (empty symbols). 1a, ▲△ C-5, ●○ C-30. 1b, ◆◇ M-4,
●○ M-8.

Hg-porosimetric measurements indicate that the secondary pore volume (pores from
75 to 75.000 Å) does not change significantly after the pretreatments (0.23-0.25
for chabazite and 0.35-0.38 cm^3/g for mordenite). The effect of reduction was
the same for silver mordenite catalysts (Fig. 1b).

X-ray diagrams of all the catalysts tested showed that the zeolite crystalline
structure was preserved during preparation and reduction.

The average crystallite size evaluated from oxygen chemisorption is given in
Table 1. They are in the range from 50 to 200 Å depending on both the silver content and the nature of zeolite. The crystallite sizes estimated from X-ray line
broadening and from electron micrographs (as in Fig. 2) were somewhat higher.
This overestimation seems to be attributed to a broad distribution of silver crystallite size with very small particles that can not be detected by X-ray. Preliminary observations carried out at higher magnification seem to confirm this hypothesis, showing silver particles ranging from some ten to several Angströms.

Results of catalytic tests on the silver-chabazite series are shown in Fig. 3
in terms of ethylene total conversion as a function of time. They exhibit high
initial activity that decreases to different conversion values depending on silver content, reaching steady conditions in 3 - 4 hours. At that time ethylene
conversion ranges from 10 to 50%. The selectivity of this series of catalysts in-

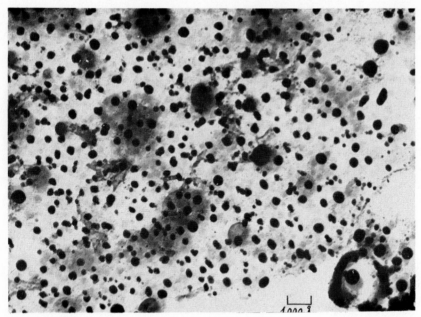

Fig.2. Electron micrograph of the sample C-20.

creases with decreasing ethylene conversion reaching at steady state low values for C-5 and C-10 and rather high values(55-60%) for C-20 and C-30 (Fig.5). This change in activity and selectivity was exhibited by the silver mordenites, too, (Fig. 4 and 5) but these catalysts reached steady values in lower time. As the pretreatment was the same for all the catalysts, it seems that the nature of zeolite can affect the change of the active surface area or the formation of Ag_xO_y films, that have been suggested as responsible of this effect. Moreover the comparison of catalytic performance of the two series shows that, at the same silver content, mordenite-based catalysts are more selective and as active as chabazite-based catalysts, except that M-8 is more active and selective. In any case both total ethylene conversion and selectivity to ethylene oxide increase by increasing silver percentage in the catalysts.

The increase in ethylene conversion with silver content does not seem due to a higher silver surface area: the apparent specific rate of both ethylene partial and total oxidation, evaluated from conversion data, increases by increasing silver content. Moreover selectivity to ethylene oxide is strongly dependent on silver content. One possible explanation of this effect has been suggested by Gior-

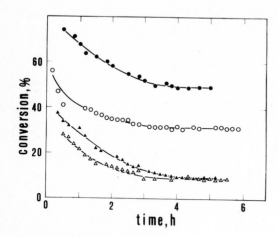

Fig. 3. Ethylene total conversion.(△ C-5, ▲ C-10, ○ C-20, ● C-30).

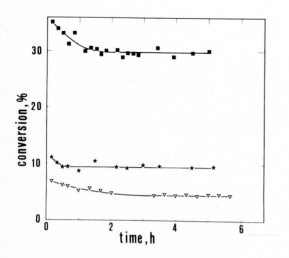

Fig. 4. Ethylene total conversion. (▽ M-2, ★ M-4, ■ M-8).

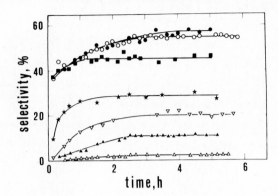

Fig. 5. Selectivity to ethylene oxide.($\triangle$ C-5, $\blacktriangle$ C-10, $\circ$ C-20, $\bullet$ C-30, $\triangledown$ M-2, $\star$ M-4, $\blacksquare$ M-8)

dano (ref. 5.) He proposes that the free zeolite surface has a negative effect on the selectivity, by degradating the ethylene oxide formed on the silver crystallites and migrated over the zeolite surface. We tested the catalytic activity of chabazite and mordenite used as supports: they resulted catalytically inert towards ethylene oxidation whentested in the same conditions of the catalysts, while by feeding a mixture of O_2 and C_2H_4O this latter was isomerized to acetaldehyde at high conversion and 80-90% selectivity. In the latter case the secondary products were CO_2 and H_2O.

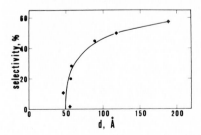

Fig. 6. Selectivity to ethylene oxide against silver crystallite size. Chabazite-series ($\blacklozenge$), Mordenite-series ($\bullet$).

As we never found acetaldehyde in the products of ethylene oxidation, the low selectivity exhibited by the catalysts at low silver content, that is due to the total oxidation of ethylene, can not be attributed to the free zeolite surface. The change of selectivity can be correlated to the silver crystallite size, supporting the hypothesis that ethylene epoxidation is a structure sensitive reaction over silver catalysts. The correlation, shown in Fig.6 indicates a large change in selectivity in the size range 50-100 Å. Similar correlations were reported by Wu and Harriott (ref.9) and by Jarjoui et al. (ref.10) for different size ranges. In the range of sizes lower than 100 Å it should be expected that crystallite size effects can be significant, resulting in different binding energies between surface atoms and between bulk atoms, and therefore, affecting the specific catalytic activity (ref. 8). Other effects related to the morphology of the crystallites (ref.8) could be involved.

Different chemical and physical properties of chabazite and mordenite could affect the silver particle size, at the same silver content, by influencing the rate of sintering of silver (ref. 11).

REFERENCES

1 X. E. Verykios, F. P. Stein and R. W. Coughlin, Catal.Rev.-Sci.Eng.22 (1980)197
2 F. Pignataro, Chim. Ind.(Milan), 62 (1980) 429.
3 P. Harriott, J. Catal., 21 (1971) 56-65.
4 N. Giordano, J. C. Bart and R. Maggiore, Z.Phys.Chem.Neue Folge 124 (1981) 97.
5 N. Giordano, J. C. Bart and R. Maggiore, Z.Phys.Chem.Neue Folge 125 (1981) 1.
6 P. Ciambelli, C.Porcelli, R.Valentino, Proceedings of the Fifth International Conference on Zeolites, L.V.C. Rees Ed.,Heyden, London, 1980, 119-128.
7 J. Papp, J. Valyon and E. Czàràn, Magy.Kêm.Folyòirat, 81 (1975) 442.
8 X. E. Verykios, F. P. Stein and R. W. Coughlin, J. Catal., 66 (1980) 368-382.
9 J. C. Wu and P. Harriott, J. Catal., 39 (1975) 395-402.
10 M. Jarjoui, B. Moraweck, P. Gravelle and S. J. Teichner, J.Chim.Phys., 75 (1978) 1060-1078.
11 M. Riassian, D. L. Trimm and P.M. Williams, J. Catal., 46 (1977) 82.

ACKNOWLEDGMENT

This work was supported in part by Research National Council (Progetto Finalizzato Chimica Fine e Secondaria).